The Final Journey
of the Saturn V

The Final Journey
of the Saturn V

Andrew R. Thomas
and Paul N. Thomarios

Ringtaw Books
Akron, Ohio

All rights reserved • First Edition 2012 • Manufactured in the United
States of America. • All inquiries and permission requests should be
addressed to the Publisher, The University of Akron Press, Akron, Ohio
44325-1703.

16   15   14   13   12   5   4   3   2   1

LIBRARY OF CONGRESS CATALOGING-IN-PUBLICATION DATA
Thomas, Andrew R.
  The final journey of the Saturn V / Andrew Thomas and Paul Thomarios.
     p. cm.
  ISBN 978-1-931968-99-7 (hbk. : alk. paper)
  1. Saturn launch vehicles—History. 2. Saturn Project (U.S.)—History. 3.
Project Apollo (U.S.)—History. I. Thomarios, Paul. II. Title. III. Title.
  TL781.5.S3T56 2011
  629.45'50973—dc23
                                    2011039576

The paper used in this publication meets the minimum requirements of
American National Standard for Information Sciences—Permanence of
Paper for Printed Library Materials, ANSI z39.48–1984. ∞

Cover: Painting by Gary Hagen, design by Amy Freels. *The Final Journey*
was designed and typeset by Amy Freels and Zac Bettendorf, and is set in
Minion and Albertus, printed on sixty-pound natural, and bound by
BookMasters of Ashland, Ohio.

Contents

For Alana—may you always dream great dreams.
Andrew R. Thomas

For my parents, Nickitas and Stilyani; my children,
Nickitas, Sarah, Adam, Emily; my third grade teacher
and member of the "Mercury 13," Jean Hixson
Paul N. Thomarios

Foreword

Distance and time tell us how far things are apart. On December 7, 1972, I left the Earth on top of a mighty and powerful Saturn V rocket. Four days, fourteen hours, twenty-two minutes, and eleven seconds later, I landed on the moon. The distance covered in that time was almost 236,000 miles. The Saturn V that carried Ronald Evans, Harrison Schmitt, and myself was the final Apollo mission. It closed the circle on President Kennedy's audacious 1961 goal of sending a man to the moon and returning him safely to the Earth. Twelve Americans walked on the lunar surface. My footprints are the last ones there. They are testament to that moment in history when human beings actually lived on another world.

The final journey of the Saturn V rocket that is on display at Kennedy Space Center was a lot different. It covered only 1.9 miles and took more than 20 years. After the cancellation of the moon missions in the early 1970s, the rocket, which was supposed to be Apollo 18, was instead laid out in the parking lot in front of the Vehicle Assembly Building—at that time the largest structure in the world. The rocket endured Florida's harsh sun, humidity, and hurricanes—but just barely. Occasionally, a coat of paint would be slapped on to keep it presentable to the visitors on the bus tour and cover up the mold and mildew. Still, it was rotting from the inside and out.

Fortunately in 1995, under the leadership of the Smithsonian Institution, a plan was put in place to restore and preserve the rocket to its original condition, and house it in the new Apollo/Saturn V Center. Selected to do the work was Paul Thomarios—the son of Greek immigrants. In May 1996, Thomarios completed the project and the refurbished Saturn V made the journey from the parking lot to its permanent home, where it continues to dazzle more than 1.5 million visitors at Kennedy Space Center each year.

This book is ultimately a celebration of the Saturn V and the indomitable strength of the human spirit. It details in simple language the rocket's creation, birth, life, death, and resurrection, so that future generations will never forget what was accomplished in the 1960s and '70s, when the courage, determination, intelligence, dedication, and slide rules of nearly 400,000 Americans were harnessed towards a single ambition: the greatest journey ever undertaken by humankind.

—Gene Cernan, Commander of Apollo 17

Acknowledgments

The creation of a book requires contributions from a great number of people, all of whom go out of their way to support the ideas and aspirations of the authors. Together, we'd first like to thank Thomas Bacher, our editor, who believed in this project from the outset and continued to challenge us to make it better than either one of us thought it could be. We'd also like to thank Amy Freels, Carol Slatter, and the many students at the Press who worked on the book. Bob Rogers of BRC, Inc. was vital in clarifying how the Apollo/Saturn V Center at Kennedy Space Center came into being. His company's work in making this vision a reality is testament to Bob's steadfastness and world-class creativity. Andrea Farmer at Delaware North Corporation, Wendy Schweiger at Sherwin Williams, and Professor Roy Hartfield of Auburn University were instrumental in helping to connect so many of the dots of this project. Karen Nelsen was more than helpful in providing her feedback and editorial experience throughout. Without Karen, this book would not have happened.

Andrew R. Thomas

I'd like to recognize the unwavering support of my mentor Tim Wilkinson in supporting my writing and academic career. My wife, Jackie, and children, Paul Bryan and Alana, were always excited as I talked endlessly about this book for years. I think my first inkling of the greatness of the Saturn V came from the Florida family vacation I took with my parents to Kennedy Space Center in 1976. Thanks Mom and Dad for this, and so much more.

Paul N. Thomarios

I'd like to thank all those that helped during the project, Carol Cavanaugh (NASA-KSC), Larry Mauk (NASA-KSC), Frank Winter (NSAM), Alan A. Needell (NSAM), Al Bachmeier (NSAM), Scott Wirz (NSAM), Bayne Rector (NSAM), Dallas Finch (Sherwin-Williams), and all the others that contributed time/efforts; my parents that taught me "never to quit;" and, my dedicated employees, who believe in me.

There are two photographs in my office of the "Wonder Women" in my life. One is of my mother. The other is of Jean Hixson—my third grade teacher during 1957. Miss Hixson was born in 1922 in Hoopeston, Illinois, the "Sweet Corn Capital of the World." At the age of 16, she persuaded her parents—her father was an insurance agent, not a risk-taking profession—to let her start flying lessons. At 18, she earned her private pilot's license. America entered World War II the following year.

My Story

Paul N. Thomarios

In 1943, Miss Hixson joined the WASPs (Women Air Force Service Pilots). Her first duty assignment was at Douglas Air Force Base in Arizona, where she towed targets for live gunnery practice, ferried aircraft domestically and overseas, and trained pilots. Later, she flew B-25 bombers over the desert at night to test their navigation systems. After the war, she took a job as a flight instructor in Akron. In her off hours, she attended the University of Akron and earned a Master's Degree in Elementary and Secondary Education.

Early in my third-grade school year, we had a substitute teacher. My classmates and I thought Miss Hixson might be sick or something even worse. The next day, the principal proudly announced to us that Miss Hixson had just become the second woman in history to break the sound barrier.

That same year, the world was turned upside down when the Soviet Union put Sputnik into space. Miss Hixson took us outside to see the shiny man-made object as it orbited overhead. Her passion for aviation and her continual insistence that the country's success was a result of America's spirit and determination were contagious. I still clearly remember Miss Hixson's lessons on how Americans could do anything once they put their minds to it.

In 1959, Hixson won a National Education Association award as "a teacher who had made outstanding use of travel and aviation experience in her classroom." Also that year, Miss Hixson was chosen to be one of America's "Mercury 13," a group of seven men and six women who had qualified for and completed astronaut training. Unfortunately, due to a later NASA decision to use only test pilots, a male-only club, Miss Hixson and her colleagues never got a chance to go into space. Miss Hixson taught in the Akron public schools for another twenty years, winning countless teaching awards and touching the lives of her students. Miss Hixson launched me forward and taught me lessons I've never lost.

Chapter 1

The Greatest Ever, Really?

Intitum didium facti (The start is half the deed).
—Roman dictum

Adjectives don't cost much. Watch a game or an awards show and count how many times superlatives are used—a hall-of-fame catch, a song for the ages, or a legendary performance. As the saying goes, "Talk is cheap." With the advent of blogs and other instant means of communication, talk is everywhere and nonstop. The here and now supersedes the historical record.

Does our hyperbole about current achievements blind us to real greatness? In some situations, the answer is a very resounding "yes." While comparing achievement from generation to generation is like asking if beauty is in the eye of the beholder, some achievements stand the test of time. Technological magnificence can be illustrated by the Great Wall of China, the Egyptian pyramids, the printing press, and the personal computer. All were significant innovations that altered culture and civilization.

Not long ago in the span of human existence, the United States built and launched the Saturn V rocket. The Saturn V was, and still is, the largest object to leave the surface of the Earth. At 363 feet in height, or over 30 stories tall, the rocket weighed 6.3 million pounds, about the weight of 1,600 automobiles or 50 Boeing 747s. In 2010, the Saturn V was taller than any building in Alaska, Delaware, Idaho, Kansas, Maine, Mississippi, Montana, New Hampshire, New Mexico, North Dakota, South Carolina, South Dakota, Vermont, West Virginia, and Wyoming.

The rocket created the loudest sound made by human hands, other than the cacophony generated by a nuclear explosion. The only natural sound on record to exceed the decibel level of the Saturn V engines was the fall of the Great Siberian Meteorite in 1883. Small earthquakes, as high as 4.6 on the Richter scale, were registered across North America when the first Saturn V launched from Florida in November 1967.

The five rocket engines of the Saturn V's first stage were the most powerful ever built. The combination of the rocket's weight and gravity's resistance required 7.7 million pounds of force to launch the rocket and its payload into orbit. By comparison, getting a jumbo jet into the air requires only 200,000 pounds of thrust.

To house the Saturn V, NASA built the Vehicle Assembly Building (VAB) at Kennedy Space Center, which remains one of the world's larger buildings, covering almost eight acres. The VAB's four mammoth doors, 456 feet in height, are the largest ever made. The VAB had to be constructed in three stages and is large enough to hold up to four complete Saturn Vs at one time. Yankee Stadium or the Rose Bowl could fit on the VAB's roof. The structure is rumored to have its own unique weather patterns.

NASA and its corporate partners built fifteen Saturn V rockets. Thirteen went into space. Twelve were used in the Apollo missions, ten of which carried astronauts and six of which took men to the moon. The last Saturn V to fly was used for the Skylab program in May 1973. Remarkably, every Saturn V launch was successful. Two missions suffered in-flight problems including engine cutoffs, but these were overcome, resulting in successful outcomes. The flawless launch record of the Saturn V stands without parallel in the history of human flight.

The Saturn V was the outcome of a pledge—President John F. Kennedy's pledge to conquer space by sending a human to the moon and returning him safely to Earth. At Rice University in Houston, Texas

on September 12, 1962, Kennedy shared his dream
with an audience of fifty thousand people.

The greater our knowledge increases, the greater
our ignorance unfolds. Despite the striking fact that
most of the scientists that the world has ever known
are alive and working today, despite the fact that
this Nation's own scientific manpower is doubling
every 12 years in a rate of growth more than three
times that of our population as a whole, despite that,
the vast stretches of the unknown and the unan-
swered and the unfinished still far outstrip our
collective comprehension.... No man can fully grasp
how far and how fast we have come, but condense,
if you will, the 50,000 years of man's recorded history
in a time span of but a half-century. Stated in these
terms, we know very little about the first 40 years,
except at the end of them advanced man had learned
to use the skins of animals to cover them. Then
about 10 years ago, under this standard, man
emerged from his caves to construct other kinds of
shelter. Only five years ago man learned to write
and use a cart with wheels. Christianity began less
than two years ago. The printing press came this
year, and then less than two months ago, during
this whole 50-year span of human history, the steam
engine provided a new source of power.... Newton
explored the meaning of gravity. Last month electric
lights and telephones and automobiles and airplanes
became available. Only last week did we develop
penicillin and television and nuclear power, and

now if America's new spacecraft succeeds in reaching Venus, we will have literally reached the stars before midnight tonight. This is a breathtaking pace, and such a pace cannot help but create new ills as it dispels old, new ignorance, new problems, new dangers. Surely the opening vistas of space promise high costs and hardships, as well as high reward.[1]

The early sixties were a heyday of hope. Technology was bringing changes to society at a record pace. A new generation was listening to the beats of rock-n-roll. Vaccines were eradicating diseases like polio. Satellites were being launched into space. In the early years of the decade, Americans could have "a meal in a minute," "live better electrically," and "fly the friendly skies." JFK was convinced that the history of the United States was one of continual achievement and that "man, in his quest for knowledge and progress, is determined and cannot be deterred."[2]

JFK's voice was full of optimism, bursting with adventure. It was time to mount a great quest, an awesome challenge.

The exploration of space will go ahead, whether we join in it or not, and it is one of the great adventures of all time, and no nation which expects to be the leader of other nations can expect to stay behind in the race for space. Those who came before us made certain that this country rode the first waves of the industrial revolutions, the first waves of modern invention, and the first wave of nuclear power, and this generation does not intend to founder in the backwash of the coming age of space. We mean to be a part of it—we mean to lead it.... Yet the vows of this Nation can only be fulfilled if we in this Nation are first, and, therefore, we intend to be first. In short, our leadership in science and in industry, our hopes for peace and security, our obligations to ourselves as well as others, all require us to make this effort, to solve these mysteries, to solve them for the good of all men, and to become the world's leading space-faring nation. We set sail on this new sea because there is new knowledge to be gained, and new rights to be won, and they must be won and used for the progress of all people. For space science, like nuclear science and all technology, has no conscience of its own. Whether it will become a force for good or ill depends on man, and only if the United States occupies a position of pre-eminence can we help decide whether this new ocean will be a sea of peace or a new terrifying theater of war.[3]

Kennedy wished to discourage the naysayers and timid. The United States was a country of doers. Given a goal, American ingenuity would win out.

But why, some say, the moon? Why choose this as our goal? And they may well ask why climb the highest mountain? Why, 35 years ago, fly the Atlantic? Why does Rice play Texas? We choose to go to the moon. We choose to go to the moon in this decade and do the other things, not because they are easy, but because they are hard, because that

goal will serve to organize and measure the best of
our energies and skills, because that challenge is
one that we are willing to accept, one we are unwill-
ing to postpone, and one which we intend to win,
and the others, too.[4]

The stage was set for a giant leap for mankind.
Questions, however, still remained. Ambitions filled
with promises were as prevalent as northern fields
covered with winter snow, but uncertainties were
everywhere. In 1960, two commercial passenger planes
had crashed over New York City in the worst aviation
disaster of the era, and Kennedy wanted the country
to go to the moon? In 1961, a Sabena flight crashed
in Belgium, killing all passengers, including the entire
eighteen-member US Figure Skating Team, and
Kennedy was aiming for the moon? Thirty-four passengers were killed on a flight to Miami. A crash near
Richmond, Virginia killed forty-eight passengers. A
crash near Montego Bay took thirty-seven lives.

In spite of the problems related to getting from one
point to the other on the surface of the planet, Kennedy
was firm in his conviction.

But if I were to say, my fellow citizens, that we shall
send to the moon, 240,000 miles away from the
control station in Houston, a giant rocket more than
300 feet tall, the length of this football field, made
of new metal alloys, some of which have not yet
been invented, capable of standing heat and stress-
es several times more than have ever been experi-

enced, fitted together with a precision better than the finest watch, carrying all the equipment needed for propulsion, guidance, control, communications, food and survival, on an untried mission, to an unknown celestial body, and then return it safely to earth, re-entering the atmosphere at speeds of over 25,000 miles per hour, causing heat about half that of the temperature of the sun and do all this, and do it right, and do it first before this decade is out—then we must be bold. However, I think we're going to do it, and I think that we must pay what needs to be paid. And this will be done in the decade of the sixties.[5]

Resolute? Yes. Momentous? Undoubtedly. Possible? As Kennedy put it, "Many years ago the great British explorer George Mallory, who was to die on Mount Everest, was asked why did he want to climb it. He said, 'Because it is there.' Well, space is there, and we're going to climb it, and the moon and the planets are there, and new hopes for knowledge and peace are there. And, therefore, as we set sail we ask God's blessing on the most hazardous and dangerous and *greatest* adventure on which man has ever embarked."[6]

---

1. President John F. Kennedy, "Speech to Rice University on the Space Effort", September 12, 1962.
2. Ibid.
3. Ibid.
4. Ibid.
5. Ibid.
6. Ibid.

The Roman historian Plutarch observed that any glory we might possess ultimately belongs to our ancestors. I am the product of all the people who have come before me. First and foremost are my parents.

My Story

Paul N. Thomarios

Like so many Americans, my parents were immigrants. Their story is the history of our great nation—suffering, sacrifice, and hard work. When success came, the fruit was very sweet.

The immigrant odyssey is daunting. Imagine leaving everything behind to start anew in a foreign land. Even though current rhetoric about America as a "melting pot" has devolved into saving ourselves by putting up fences, Richard Herman, coauthor of *Immigrant, Inc.*, understands why the influx of peoples to the United States is vital to this country. "[I]mmigrants are more likely to start a business, invent something, earn an advanced degree, and have intimate knowledge of global markets than "native-born" Americans." Perhaps we all need to travel a little more.

Before becoming immigrants, my parents were refugees, lived under Nazi-occupation, and became refugees again. Nobody gave them anything and almost every promise ever made to them was broken. Remarkably, my parents were never bitter or critical of the hand that life dealt them. They rarely, if ever, complained.

Their goal was to build a better life with determination, know-how, extra hours, and improvement. All my parents wanted was a chance. They had learned how to survive under adverse conditions. America gave them their chance and they took it.

Chapter 2

Audacity of the Highest Order

The first quality that is needed is audacity.
—Winston Churchill

Weighing only 180 pounds and roughly the size of a basketball, Sputnik was launched in 1957 and changed the world. The satellite orbited the Earth every 96 minutes, and the launch inflicted a psychological blow on the United States. The Soviets, professing a doctrine which was the antithesis of democracy, had blasted out of the starting blocks and were sprinting ahead in the space race. Soviet engineers, in a move as much public relations as scientific importance, had built Sputnik with four antennas trailing behind it and had made the device out of shiny, polished metal so it could be seen with the naked eye 175 miles up in the sky. Even though Sputnik broadcast an insignificant "beep-beep" pattern which could be picked up by amateur radio operators around the globe, the launch proved the Soviets were capable of major scientific advances. For most Americans, the launch was a sober warning that before long the Soviets would be spying into Americans' backyards.

If Orson Welles' 1938 broadcast of *War of the Worlds* was an indication of how a phony invasion could panic Americans, the Sputnik launch was a very real symbol that the Soviets had the capability to launch missiles into space and at targets in the United States. After all, the Cold War was being waged daily. The world was divided into two blocs—the Soviet-dominated eastern bloc and the United States-dominated western bloc. The tensions between the two blocs would eventually lead to the walling off of Berlin in the late summer of 1961 and the Cuban Missile Crisis in October 1962, bringing the world to the brink of nuclear war. US Civil Defense videos of the era warned of nuclear attacks and outlined evacuation plans. School children were taught to "duck and cover" in case of bombings. Because both nations possessed increasingly larger numbers of more accurate and devastating nuclear weapons—pointed at the other's cities and towns—there was genuine fear military action instigated by one of the countries could destroy the whole world.

Thankfully, the Cold War wasn't fought directly. It was fought with words and will. It was contested in political and economic realms. The Cold War was contested with financial and military support for Soviet and United States' allies. The Olympic Games became a proving ground for the competing doctrines. The battle for space was no different.

During the early years of the Space Race, as the contest between the two superpowers came to be called, success was marked by headline-making "firsts." At the outset, the Soviet Union was taking home the trophies.

- In 1957, Sputnik was the first satellite launched.
- In 1959, Luna 2, a Soviet space probe, made it to the moon.
- In 1961, Soviet cosmonaut Yuri Gagarin became the first person to orbit the Earth.
- In 1963, Valentina Tereshkova became the first woman in space.
- In 1965, Alexei Leonov becomes the first person to walk in space.

How could the United States be losing? Was the Soviet system better at producing results? American pride and security were on the line. A country that had persevered in World War II and designed an apparatus to end the war was now being left behind at a critical period in world history. Ironically, when President Kennedy made his 1961 speech before Congress announcing the goal to put an American on the moon, only one American had ever flown in space, Alan Shepard, and his flight had taken place only two weeks before Kennedy's speech. How could Shepard's Freedom 7 flight, lasting a little more than fifteen minutes, be the first step toward moon exploration?

As the United States looked to explore space, serious questions were commonplace, but answers seemed as apparent as stars on a cloudy night. Many of the most basic issues needed to be resolved. A few years earlier, engineer Ernst Stuhlinger briefed US government officials about the plethora of unknowns surrounding manned space flight.

> What happens, for instance, to metals, plastics, sealants, insulators, lubricants, moving parts, flexible parts, surfaces, coatings, and liquids in outer space? How could we guard men and materials from the dangers of radiation, meteorites, extreme temperatures, corrosion possibilities, and weightlessness? What kinds of test objectives, in what order and how soon, should be established? We are of the opinion that if we fail to come up with answers and solutions to these problems, then our entire space program may come to a dead end.[1]

These questions constituted only a few of the details the American scientists hadn't resolved. Obviously, the biggest question revolved around the technology to get men to the moon. Once there, however, how could those astronauts be brought back safely? Questions led to more questions. Kennedy had laid out the challenge, and America was energized to win the race to the moon, but American know-how would have to leap forward quickly to ensure the result.

A Brief History of Rockets

Not all grand aspirations lead to successful outcomes. In fact, failure in most human endeavors is the norm. We're good at trial and error, and understand if we don't succeed, we'll have to try again and again and again. Eventual success might take decades and comes because a myriad of disparate factors have fallen into place. As we make attempts, attempts at climbing Mt. Everest, for example, many dynamics are beyond our control. Building and launching the Saturn V rocket, the "means" to the greater "end," was a spectacular accomplishment, especially in the face of so many unknowns.

Humans are marvelously fit for life on Earth. Our bodies have evolved over millions of years and have acclimated to the protection of the atmosphere that separates us from outer space. The race to the moon, however, wasn't one that would allow evolutionary processes. Kennedy's goal needed quicker catalysts. In place of natural development, a myriad of challenges would have to be solved with human innovations and technological advances. The Saturn V rocket would become the greatest machine ever built. Fortunately, American scientists and researchers didn't have to start from scratch.

It is widely believed the earliest rockets originated in China, not long after the invention of gunpowder somewhere around the tenth century. Over the next few centuries, the use of rockets in the form of artillery shells expanded across China and eventually, by the fourteenth century, the technology was exported to Europe. Rockets were used almost exclusively for military purposes during the first five hundred years of their existence.

Jules Verne, the French novelist, opened up the possibilities for rocket use by describing space flight in his 1865 novel *From the Earth to the Moon*. Verne's "moon gun" was eerily similar to the Saturn V, which would be developed nearly a century later. Indeed, much of what Verne predicted in his novel would come true. He indicated "the United States would launch the first manned vehicle to circumnavigate the moon." Verne thought this would be the case because nothing "can astound an American. It has often been asserted that the word 'impossible' is a French one. People have evidently been deceived by the dictionary. In America, all is easy, all is simple; and as for mechanical difficulties, they are overcome before they arise. A thing with them is no sooner said than done."[2]

Verne also raised the more technical questions for going to the moon and he provided remarkably accurate answers. On the question as to whether it was even possible to send a projectile to the moon, Verne's opinion was unequivocal:

> Yes; provided it [a projectile] possess an initial velocity of 1,200 yards per second; calculations prove that to be sufficient. In proportion as we recede from the earth the action of gravitation diminishes in the inverse ratio of the square of the distance; that is to say, at three times a given distance the action is nine times less. Consequently, the weight of a shot will decrease, and will become reduced to zero at the instant that the attraction of the moon exactly counterpoises that of the Earth; that is to say at 47/52 of its passage. At that instant the projectile will have no weight whatever; and, if it passes that point, it will fall into the moon by the sole effect of the lunar attraction. The theoretical possibility of the experiment is therefore absolutely demonstrated; its success must depend upon the power of the engine employed.[3]

Many of Verne's predictions in *From the Earth to the Moon* were well ahead of the times.

- Verne named the cannon used to launch his spacecraft "Columbiad." The command module of the Apollo 11 moon mission was called *Columbia.*

- After considering twelve possible launch sites in Texas and Florida in his novel, Verne selected Stone Hill, south of Tampa, Florida. One hundred years later, NASA considered seven launch sites and chose Merritt Island, Florida.

- Verne's spacecraft was launched in December, from 27° 7' N, 82° 9' W. After a journey of 242 hours, 31 minutes, including 48 hours in lunar orbit, the spacecraft splashed down in the Pacific Ocean at 20° 7' N, 118° 39' W, and was recovered by the US Navy vessel *Susquehanna*. The crew of Apollo 8, the first manned Saturn V flight, was launched in December 100 years later, from latitude 28° 27' N, 80° 36' W (132 miles or 213 km from Verne's site). After a journey of 147 hours, 1 minute, including 20 hours, 10 minutes in lunar orbit. The spacecraft splashed down in the Pacific Ocean at (8° 10' N, 165° 00' W) and was recovered by the US Navy vessel *Hornet*.

The Rocket Pioneers

Verne deserves much of the credit for inspiring early rocket pioneers, especially three men whose work laid the foundation for the creation of the Saturn V.

In the late 1800s, Russian schoolteacher Konstantin Tsiolkovsky proved mathematically that rockets could propel objects and human beings into space. Tsiolkovsky is considered by many to be the father of modern spaceflight. In 1903, he published his truly groundbreaking mathematical equation, known today simply as "The Ideal Rocket Equation." It continues to serve as the foundation for rocket propulsion.[4]

Tsiolkovsky realized that rockets needed to produce their own thrust in order to fly, whether in space or when leaving the gravity of the Earth. In space and in the upper atmosphere there is no air. Therefore, a rocket must carry both its oxidizer and fuel. The space shuttle does not stay in space because of lift from its wings but because of orbital mechanics related to its speed. Space is nearly a vacuum. Without air, there is no lift generated by the wings. Tsiolkovsky based his ideas on Newton's Third Law of Motion—"For every action there is an equal and opposite reaction." When a rocket uses its own engines and forces hot gas down towards the ground, the rocket will move in the opposite direction: up! A rocket launch becomes a "tug-of-war" between the rocket's propulsion and gravity. If the rocket is able to push harder than gravity pulls, we have a liftoff. This pushing force is called thrust.[5]

The next major steps along the evolutionary path for rocketry were undertaken by the American, Robert

Goddard, who was also inspired by Verne as a kid. In 1914, Goddard was awarded US Patent Number 1,102,653 for a multistage rocket. The patent's description begins to foretell the future of rocketry.

> This invention relates to a rocket apparatus and particularly to a form of such apparatus adapted to transport photographic or other recording instruments to extreme heights ... is performed with great efficiency whereby the velocity and range of flight are greatly increased.[6]

In 1919, the Smithsonian Institution published Goddard's benchmark book, *A Method of Reaching Extreme Altitudes*. Regarded as Goddard's most important work, the book is considered to be one of the more influential books ever written about rocket science. In the text, Goddard explains that at a velocity of 6.95 miles per second, without air resistance, an object can leave Earth's gravity and "escape to infinity." He also noted "that one or more rockets, really copies in miniature of the larger primary rocket, should be used if the most extreme altitudes are to be reached."[7] In other words, the rocket had to be built of several stages, each of which could launch the rocket further along its journey.

On March 16, 1926, Goddard launched his first liquid-fueled rocket, a liquid oxygen and gasoline vehicle that rose 184 feet in 2.5 seconds. This event heralded the modern age of rocketry. He continued to experiment with rockets and propellants for the rest of his life. From 1930 to 1941, he launched rockets of increasing complexity and capability. He developed systems for steering a rocket in flight by using a rudder-like device to deflect the gaseous exhaust. He also used gyroscopes to keep the rocket headed in the proper direction.

The culmination of this effort was a successful 1941 launch of a rocket to an altitude of nine thousand feet. Later that year, Goddard joined the US Navy and spent the duration of World War II developing a jet-assisted takeoff (JATO) rocket to shorten the distance required for heavy aircraft launches. Some of this work led to the development of the "throttle-able" Curtiss-Wright XLR25-CW-1 rocket engine, which later powered the *Bell X-2* research airplane that helped break the sound barrier in 1947. Goddard did not live to see this; he died in Baltimore, Maryland, on August 10, 1945.

An ardent follower of Goddard's work, Wernher von Braun was one of the most important rocket developers and champions of space exploration between the 1930s and the 1970s. As a youth, like Tsiolkovsky and Goddard, von Braun also became enamored with the possibilities of space exploration by reading Jules Verne, which prompted him to master calculus and trigonometry so he could understand the physics of rocketry.

In 1932, to further his desire to build large and capable rockets, von Braun went to work for the German army to develop ballistic missiles. While engaged in this work, von Braun received a Ph.D. in physics. Von Braun is well-known as the leader of what has been called the "rocket team," which developed the V–2 ballistic missile for the Nazis during World War II.[8]

The brainchild of von Braun's rocket team, operating at a secret laboratory in Peenemünde on the Baltic coast, the V–2 rocket was the precursor of rockets used in early space exploration programs of the United States and the Soviet Union. A liquid propellant missile extending some 46 feet in length and weighing 27,000 pounds, the V-2 was capable of flying at speeds in excess of 3,500 miles per hour and delivering a 2,200-pound warhead to a target 500 miles away.[9]

First flown in October 1942, the V-2 was employed against targets in Europe beginning in September 1944. By the beginning of 1945, it was obvious to von Braun that Germany would not achieve victory against the Allies and he began planning for the postwar era. Before the Allied capture of the V–2 rocket complex, von Braun engineered the surrender of five hundred of his top rocket scientists, along with plans and test vehicles, to the Americans.

As part of a US military operation called Project Paperclip, von Braun and his rocket team were

scooped up from defeated Germany and sent to Fort
Bliss, Texas. There, they worked on rockets for the
US Army, launching them at White Sands Proving
Ground, New Mexico.[10] In 1950, von Braun's team
moved to the Redstone Arsenal near Huntsville,
Alabama, where they built the Army's Jupiter ballistic missile. In 1960, von Braun's rocket development
center transferred from the Army to the newly-estab-
lished National Aeronautics and Space Administration (NASA) and received a mandate to build the
giant Saturn rockets.

While von Braun's role in the Nazi war effort was
never in doubt, as Michael Neufeld, von Braun's biog-
rapher points out, no one played a greater part in
making Kennedy's dream a reality than von Braun.[11]
Von Braun's role in the eventual moon landings was
indispensable. Von Braun's technical expertise was
without comparison, and he was also the most out-
spoken and well-known proponent of space flight
during the 1950s in the United States. Under his
tenacious leadership, the Saturn family of rockets
were developed, tested, and built.[12]

In an April 29, 1961, letter to Vice President Lyndon
Johnson, von Braun reassured the Kennedy Administration that it was quite possible to get to the moon
before the Soviets. Von Braun wrote, "We have an
excellent chance of beating the Soviets to the first
landing of a crew on the moon (including return
capability, of course). The reason is that a performance

jump by a factor 10 over their present rockets is necessary to acquire this feat. While today we do not have
such a rocket, it is unlikely the Soviets have it."[13] Von
Braun then summarized what would become Amer-
ica's national priority in the coming years: "With an
all-out crash program I think we could achieve this
objective [putting a man on the moon] in 1967/68."[14]

---

1. "ABMA Presentation to NACA," 1957, pp. 129–30, Army
Ballistic Missile Agency Archives, Washington, DC.
2. Jules Verne, *From the Earth to the Moon* (New York: Barnes
& Noble Books, 2005), p. 328.
3. Ibid., p. 329.
4. Brigham Narins, ed., *Notable Scientists from 1900 to the
Present #5* (Farmington Hills, MI: Gale Group, 2001), pp. 2256–
58.
5. The Tsiolkovsky rocket equation, or ideal rocket equation,
is a mathematical equation that relates the delta-v with the
effective exhaust velocity and the initial and end mass of a
rocket. It considers the principle of a rocket: a device that can
apply an acceleration to itself (a thrust) by expelling part of its
mass with high speed in the opposite direction, due to the
conservation of momentum. For any such maneuver, or journey
involving a number of such maneuvers, the formula is

$$v_e = I_{sp} \cdot g_0$$

where:al total mass, including propellant, in kg (or lb)
m_1 is the final total mass in kg (or lb)
v_e is the effective exhaust velocity in m/s or (ft/s) or

$$\Delta v = v_e \ln \frac{m_0}{m_1}$$

Δv is the delta-v in m/s (or ft/s).
6. Robert H. Goddard. Rocket Apparatus. US Patent 1,102,653,
filed October 1, 1912, and issued July 7, 1914.

7. Robert Goddard, *A Method of Reaching Extreme Altitudes,* (Washington, DC: Smithsonian Institution, 1919), p. 57.

8. "Biography of Wernher von Braun," Marshall Space Flight Center, accessed August 10, 2009, http://history.msfc.nasa.gov/ vonbraun/bio.html.

9. Ibid.

10. Ibid.

11. Michael Neufeld, *von Braun: Dreamer of Space, Engineer of War* (New York: Vintage Books, 2008), p. 5. Neufeld's book is the definitive work on von Braun and the efforts of the NASA Marshall Space Flight Center team he lead before and during the Apollo program.

12. Neufeld, pp. 5–6.

13. Werner von Braun to Lyndon B. Johnson, April 29, 1961, NASA Historical Collection, NASA Headquarters, Washington, DC.

14. Ibid.

My father, Nickitas, survived the
killing of the Greeks during and after
World War I. Before the war, 3 million
Greeks resided in Turkish territory.
Prior to 1915, 1.5 million of them
were killed or dispersed; after 1915,

My Story

Paul N. Thomarios

750 thousand were killed or dispersed. Some were sent to
arid plateaus and found little food or water. The survivors,
my parents included, were shipped to Greece and became
refugees. As the May 5, 1915, *New York Times* put it, "To
tear hundreds of thousands of persons from the places they
had regarded for generations as their rightful homes was
considered more humane than to leave them there under
the threat of death." My mother and father ended up in the
area around Lavrion, the chief mining region of Greece.

The history of mining in that area goes back to the Era
of Pericles, during 5th Century B.C. As the eminent historian Will Durant noted in *The History of Civilization*, "We
forget that except for machinery and our religions, there is
hardly anything in our culture today that does not come
from Greece. Schools, gymnasiums, arithmetic, geometry,
history, rhetoric, physics, biology, anatomy, hygiene, therapy,
cosmetics, poetry, music, tragedy, comedy, philosophy,
theology, agnosticism, skepticism, stoicism, Epicureanism,
ethics, politics, idealism, philanthropy, cynicism, tyranny,
plutocracy, and democracy: these are all Greek words for
cultural forms seldom originated, but in many cases first
matured by the abounding energy of the Greeks." The silver
and gold of Lavrion financed the building of the Acropolis,
the Parthenon, and other structures that founded Western
civilization.

My father was 10 years old when he arrived in Lavrion.
At the age of 12, he went to work in the Merchant Marines.

For the next six years, he was a fireman and he shoveled coal into the furnaces on steam ships. When my father turned 18, he had two options—join the Army or the Navy. Since my father was already a sailor, he chose the Navy. He served the mandatory three years and then returned to the Merchant Marines. Unmarried and 27 years old, he went to my grandmother and said, "It's time for me to get married. Do you know anybody around here?" The courtship rules were different then.

Chapter 3

The Game Changers

Never tell people how to do things. Tell them what to do and they will surprise you with their ingenuity.
—General George S. Patton

On July 3, 1969, a massive fireball, visible from space, erupted at the Baikonur Cosmodrome. The same facility had witnessed the successful launching of Sputnik twelve years earlier, but this time the N-1 Soviet rocket trying to liftoff came crashing and burning back to Earth. The inferno destroyed the launch complex and its fifty-eight-story tower. Windows in buildings thirty miles away were completely shattered. The explosion left scars on the terrain that were evident in commercial satellite photos even twenty years later.

The N-1 was the Soviet moon rocket. It had 3 stages and rose to an overall height of 347 feet. Incredibly, the first stage had 30 engines, the second had 8, and the third had 4. The Soviets built ten N-1s. Four were test fired and failed. The 1969 disaster was a devastating blow to the Soviet lunar landing program and the remaining six N-1s were dismantled in 1974, after the Soviets gave up their dream to land men on the moon. Soviet scientists couldn't overcome the challenge of getting thirty engines in the first stage to work together.

On July 16, 1969, a massive machine carrying Neil Armstrong, Edwin "Buzz" Aldrin, and Michael Collins was readied for launch at Cape Canaveral in Florida. Armstrong, who would become the first human to walk on the moon, had a pilot's license before his driver's license and built wind tunnels in his parent's basement. As a youth, he was shy, unassertive, and not very athletic. Armstrong, however, represented in many ways the entire American program to land a man on the moon. Not only did Armstrong have a fervent zeal for aviation, he had a quiet drive and stoic decisiveness. The descriptive line in his yearbook summed it up well. "He thinks, he acts, 'tis done."[1]

Apollo 11 was successfully launched and with the words "the Eagle has landed," the command module touched down on the lunar surface. At 2:56:15 GMT on July 21, 1969, Neil Armstrong made his "one small step for man [and] one giant leap for mankind." The United

States had won the space race. The story of Armstrong's historic moment was close to twenty years in the making and involved hundreds of thousands of individuals.

In the wake of the early Soviet victories in the space race, NASA chief James Webb was faced with speeding up rocket development. Webb, knowing that similar technologies were used by ballistic missile experts, turned to Wernher von Braun's team, based in Huntsville, Alabama. Von Braun's team, along with a stable of private-sector contractors and subcontractors, had been designing and building much of America's nuclear arsenal since the end of World War II. After the formation of NASA in 1958, a significant portion of America's ballistic missile infrastructure was eventually brought into the new space agency.

The scope and magnitude of designing the Saturn V moon rocket required a vast network of American companies and universities, led by NASA. Outside contractors, prime, second, and third, were comprised of nearly twenty thousand companies, two thousand of which were prime contractors. In addition, approximately two hundred universities and technical institutes contributed to the program.

Without the vital role of the outside contractors, America would never have gotten to the moon and back. Between 1961 and 1971, 85 percent of the NASA budget was turned over to contractors.[2] In total, nearly four hundred thousand Americans worked on the Saturn V, most of them for privately-held firms and research universities. The flag of the Apollo program was carried by NASA and its astronauts, but the true heroes were the hundreds of thousands of unnamed and unheralded technicians and engineers who had the "right stuff" necessary to research and execute the undertaking. In an era devoid of personal computers, cell phones, laptops, or other handheld devices, the results of the Apollo program are all the more mind-boggling.

The Saturn Family of Rockets

The Saturn V was the latest in the line of an incredibly successful family of Saturn rockets. The Saturn family descended from the Juno super booster, which had been developed by von Braun and his team. On November 18, 1959, Saturn was transferred to NASA. The Saturn family consisted of three rockets, each designed for a specific purpose: the Saturn I, Saturn IB, and Saturn V.

The Saturn Family of Rockets

	Saturn I	Saturn IB	Saturn V
Original Name	C-1	C-1B	C-5
Height	190 feet	223 feet	363 feet
Mission	Unmanned Earth orbit	Manned Earth orbit	Manned Lunar landings
Rocket Stage 1	S-1	1-B	1-C
Rocket Stage 2	S-IV	1V-B	S-II
Rocket Stage 3	Apollo spacecraft	Apollo spacecraft	IV-B
Rocket Stage 4	n/a	n/a	Apollo spacecraft

All of the Saturns were multi-stage rockets, a feature vital for landing a man on the moon. The smallest of the Saturn family, the Saturn I was 190 feet in height, making it taller than the space shuttle when that vehicle was launched on April 12, 1981. The Saturn I, a two-stage rocket, was designed for unmanned Earth orbit, and its unique design of strapping eight engines together was a key to its success. The first Saturn I rocket was launched on October 27, 1961 and weighed 925,000 pounds. The final flight of the Saturn I took place on July 30, 1965.

The "middle child" was the Saturn IB, a two-stage rocket that orbited the Earth carrying Apollo astronauts as they trained for the moon missions. On October 11, 1968, the Saturn IB launched the first manned Apollo spacecraft, Apollo 7 and after the completion of the Apollo program, launched three missions to man the Skylab Orbital Workshop in 1973. In 1975, the IB launched the American crew for the Apollo/Soyuz Test Project, the joint US/Soviet Union docking mission. The Saturn IB had more advanced upper-stage engines than the Saturn I.

The "big brother" of the Saturns was the Saturn V. The first manned flight of the Saturn V was Apollo 8, a flight that orbited the moon in December 1968. Six manned lunar landings were fueled by Saturn V rockets; the last lunar exploration flight took place on December 7, 1972, a day that would not live in infamy because the mission highlighted the results of the Saturn launches—thirty-two launches, thirty-two successes.

Early in the Apollo program, after analysis by engineers, the decision was made to go with a three-stage rocket. A one-stage rocket would require so much fuel and engines so massive that such a craft simply would not get off the ground. A two-stage rocket, while more promising, would not be able to create enough thrust to reach the moon.

Staging, as creating separate thrusting units came to be known, is a simple notion. For the Saturn V, the process worked as follows:

> The first stage, which contained the largest engines and quantities of fuel, would provide enough velocity to get the entire rocket to a high altitude. Once the first stage's fuel was depleted, the first stage would be jettisoned and fall back to Earth. Engines on the second stage would then fire, pushing the rocket higher. Likewise after the second stage's fuel was exhausted, that stage would fall away. The third stage propelled the command and lunar modules into Earth orbit and provided the final thrust to send the payload toward the moon.

The Game Changers
The unrivaled success of the Saturn V and its predecessors was due in large part to innovations within the US ballistic missile program, innovations that

occurred years before President Kennedy's 1961 speech. In many ways, the critical enabling technology for the Saturn V program was the engine design: specifically the F-1 engines on the first stage and the liquid hydrogen-fueled J-2 engines on the second and third stages.

The "V" of the Saturn V came from the five F-1 engines on the first stage. To this day, the F-1 engines remain the largest, most powerful liquid-fueled engines ever built by human beings. The F-1 was developed in the mid-1950s by Rocketdyne Corporation in response to a request from the United States Air Force.

One of the game changers at Rocketdyne was Bob Biggs. He joined the company after a two-year stint in the Army. He was stationed in Southern California with an anti-missile defense battalion. Originally from Nashville, Bob met his wife during his time in the San Fernando Valley and decided to stay in California after leaving the Army.

A problem-solver at heart, Bob studied engineering for two years at Vanderbilt before joining the Army. In 1957, a neighbor of Bob's encouraged him to apply for a job at Rocketdyne. The company was expanding quickly in order to ramp up engine production for US ballistic missile systems. Almost immediately, Bob started working on rocket engine development. At the same, he continued to study electrical engineering at UCLA.

In 1960, Bob moved to Rocketdyne's F-1 program on the Juno missile, the forerunner of the Saturn V. When the American space program's goal became landing a man on the moon, Bob was given a new task: figure out how to ignite the F-1 engine. Over the next several years, Bob was given the resources needed to accomplish his undertaking and the outcome was stunning.

After starting, the 5 first-stage engines burned 3,357 gallons of fuel per second, the equivalent of emptying a 30,000-gallon swimming pool in less than 9 seconds! The engines churned out more power than 85 Hoover Dams or enough energy to light up New York City for 75 minutes. Igniting seconds before actual liftoff, the engine's turbo-pumps, with the power of 30 diesel locomotives, forced 15 tons of kerosene and liquid oxygen fuel per second into the thrust chambers.

Bob was present for the first few tests of the F-1 engine, which took place at Edwards Air Force Base. He also witnessed the launch of Apollo IV, the first Saturn V, on November 9, 1967. Biggs later described the launch in an article he wrote for an internal Rocketdyne publication: "Watching the rocket, the sound was unbelievably awesome from three miles away. There was an intense vibration of my rib cage due to the low frequency generated by the engines." Bob continued, "There was a feeling of euphoria which overcame me, and when I saw the rocket rise, I was substantiated by joy, pride, and awe. At that moment I was glad I chose to be an engineer."[3]

Joe Strangeland was another key Rocketdyne engineer, responsible for the structural integrity of the turbo-pumps, which fed the fuel to the monstrous F-1. Raised on a Minnesota farm, Joe was fascinated by aviation as a child. During World War II, a Boeing B-29 made a forced landing in a grass airstrip across the fence from where he grew up. The plane came in during the middle of the night and Joe remembers his neighbors shining lights on the airstrip to guide the pilot's descent. From that moment, Strangeland had the aviation bug.

Joe left home when he was a senior in high school and went to study in Rapid City, South Dakota, at the School of Mines and Technology. He graduated as a mechanical engineer in 1957 and headed west to join Rocketdyne. He began his stint at Rocketdyne in the structures department. Joe's primary responsibility was working on the engines, and specifically, the turbo-pumps.

From the first day he joined Rocketdyne, Joe recalls that "no was never used very often." People routinely, and without fanfare, uncovered a lot of problems and just solved them. In 1959, Joe started working on the turbo-pumps of the F-1. Half-joking, Joe quips, "the F-1 engine was merely a mount for his turbo-pump."[4]

At the same time he worked on the pumps for the F-1, Joe also began to work on the J-2, the engine that would propel the second and third stages higher, and ultimately, into Earth and then lunar orbit. The work was difficult because of the engines' unique use of liquid hydrogen.

The Saturn V was designed to push the Apollo spacecraft into orbit. Unlike a truck, which pulls the cargo that is located behind it, the Saturn V pushed the service, command, and lunar excursion modules through the atmosphere and into space. Liquid hydrogen fuel appealed to Saturn V's builders because of what rocket scientists had discovered about its high "specific impulse"—a basic measure of how a rocket engine performs. Compared to a similar-sized engine which used kerosene as its fuel, a liquid hydrogen engine could increase the specific impulse by 40 percent.

Specific impulse is the number of pounds of thrust produced per pound of propellant used per second, or, in other words, a rough measure of how fast the propellant is ejected out of the back of the rocket. The speed of a rocket depends on thrust compared to the rocket's weight. The faster the speed at which propellant is thrown out the back of the rocket, the faster the rocket can travel or the more cargo it can carry. Therefore, a rocket with a high specific impulse doesn't need as much fuel as a rocket with low specific impulse. The higher the specific impulse, the more push you get for the fuel that rushes out.[5]

Another game changer at Rocketdyne was Paul Coffman, a native of Southern California. He took a summer job at the company in 1955 while attending Los Angeles City College, and then accepted a part-time job at the end of the summer. After a couple of years, he began to work full-time while he completed his degree in mechanical engineering. He graduated from USC in 1959.

According to Paul, the J-2 engine was a "technology challenge." As late as 1958, reports from the National Bureau of Standards indicated that research and knowledge about liquid hydrogen as a fuel were still widely lacking. In part to close this knowledge gap, Coffman served as lead engineer on the J-2 thrust chamber assembly development, supervisor of engineering test for J-2 components and engines, and manager of J-2 engine development and flight support. In 1961, to tackle the knowledge gap, Paul moved from a test engineer in development to a manager for the J-2. He worked at Rocketdyne until 1971.[6]

For these engineers and their counterparts around America, the 1960s were a blur. Sixty- to eighty-hour work weeks were the norm. Says Paul: "Almost all of us who worked on this project were young, ignorant, and naïve. But we kept our priorities."[7] Joe Strangeland said he never worked harder in his life and was never bored. "There were no computers; no private offices. We shared everything: phones, bathrooms, wastebaskets. The only thing that mattered was the work."[8]

These three game changers are examples of the people who powered America to the moon. Of course, their stories are not unique, but a part of an American landscape filled with individuals who used ingenuity, creativity, and sheer will to accomplish a singular goal. Think of Biggs, Strangeland, and Coffman. Think of multiplying their hard work by a factor of a hundred thousand. Think of the approximately 578,000-mile round trip from the Earth to the moon. Each mile was filled with small steps taken by everyday citizens so Neil Armstrong could take his giant leap for mankind!

---

1. From the biography of Neil Armstrong as depicted at the National Aviation Hall of Fame, Dayton, Ohio.

2. Nicolas Turcat, "The link between aerospace industry and NASA during the Apollo years," *Acta Astronautica* 62 (2008): 66–70.

3. Bob Biggs, interview by Andrew R. Thomas, February 1, 2010.

4. Joe Strangeland, interview by Andrew R. Thomas, February 5, 2010.

5. Rocketdyne Corporation, "Propulsion: The Key to Moon Travel," 1964.

6. Paul Coffman, interview by Andrew R. Thomas, February 5, 2010.

7. Ibid.

8. Ibid.

I guess you could say my parent's marriage was arranged. Maybe it made the whole process easier. I'm not sure.

My Story

Paul N. Thomarios

After my father asked for his mother's help to find a wife, she said, "Yes, there is a family with a daughter that has our same history and I'm very good friends with her mother. Why don't you meet her and see if you like each other?" He did. They were engaged. In two months, they became husband and wife. My mother became pregnant quickly. My father's return to the Merchant Marines required that he go to sea for three years. After my father boarded his ship, my pregnant mother didn't see him again for almost a decade.

During my father's tour of duty, the Nazis invaded Greece. In April 1941, upon his arrival in Seattle, Washington, my father learned that Greece had fallen and was now under Nazi occupation. For the second time in his young life, my father was a man without a country.

Stuck in Seattle, with little knowledge of English and no friends, my father's only option was to reach out to a pair of uncles who had previously emigrated to Akron, Ohio. One uncle ran a small tailor shop and the other did shoe repair. My father bought a ticket, got on the train, and got off 2,400 miles later in Akron. For the next few months, he worked odd jobs and tried to assimilate to his new home. In December 1941, the Japanese attacked Pearl Harbor. My father was drafted into the United States Army, served proudly until 1946, and was discharged as a sergeant.

Chapter 4
Putting It All Together

Everything should be as simple as possible, but not
one bit simpler.
—Albert Einstein

You have to stipulate we were very, very lucky, but
we did it just the same.
—US official's comment after the 1961 launch of
Saturn's first stage.

In February of 1964, NASA's Director of Manned
Space Flight forecast that the United States would
land a man on the moon by 1968 or 1969. Considering that just two years earlier, experts in London
predicted the Soviets would win the space race,
NASA's bold statement seemed as realistic as the

moon being made of cheese. Over 3 million parts, making up 700,000 components, were needed to construct one Saturn V. How could such a feat be achieved in such a short time? While money did talk (in this particular case it spoke volumes—approximately $300 billion), funding alone wouldn't get the job done. The coordinated commitment and innovation of of nearly 400,000 people was the eventual ticket to success.

At the beginning of the space race, Soviet and US rockets were not capable of lifting the needed payloads to reach the moon. The Soviets had an early lead, since their military rockets, redesigned for space exploration, were originally built in the early 1950s to launch heavier nuclear warheads and were twice as reliable as the best in the US arsenal. The Soviet Sputnik satellite weighed a little over 184 pounds. The US Mercury capsules that, beginning in 1962, orbited the Earth weighed three thousand pounds. To get a man to the moon and back, NASA estimated that the command and lunar modules would need to weigh almost sixteen times greater than the Mercury capsules.

As NASA began to plan for the Saturn program, safety was a major concern. Technology snafus were inherent in the trial and error of rapid development. Limiting disastrous consequences was not only necessary, but vital to keep morale high and inspire further progress. In the early sixties, many Americans still remembered the 1957 Vanguard TV3 satellite launch debacle, which Lyndon Johnson called the "most humiliating failure in America's history."[1]

The Saturn rocket, rising thirty-six stories and weighing as much as a naval destroyer, was pent-up power galore. If the Saturn rocket were to explode at or close to launch, the immediate area would be hit with the force of a small atomic bomb—the equivalent of one-half a kiloton or about $\frac{1}{26}$th the force of the bomb that destroyed Hiroshima.[2] Obviously, an explosion of this magnitude would lead to many deaths and destroy the launch area completely. The subsequent investigations and second-guessing would have caused significant delays or the dismantling of the entire program.

Further, Kennedy's challenge put the spotlight squarely on the space program. While the Soviets were able to hide failures and mishaps, some of which only became apparent after the Soviet regime fell, an open press precluded the same scenario in the United States. Any major malfunction during launch or flight would have been witnessed by a global audience. Any disaster would not only have damaged US prestige, but undoubtedly would have led to Congressional hearings. Catastrophic failures in testing and launching, therefore, had to be avoided at all costs. The rocket design and development process had to be created and monitored with the highest quality assurance procedures. Ambition had to be embraced but the devil could not get into the details.

The ways and means to the moon were not fully ironed out at the end of 1962. While many in the space agency favored a lunar orbit approach that would be used to deploy a landing module to the surface, Jerome Wiesner, Kennedy's Special Assistant for Science and Technology, favored an Earth orbit rendezvous, an approach that would have two rockets circle the Earth and one would fuel the other for the moon trip, or the direct ascent to the moon approach. There was even talk of sending a man to the moon on a one-way mission with enough supplies and facilities until another mission's crew could pick him up.

In the spring of 1963, after another successful Soviet launch, members of the House Space Committee contemplated cutting the Apollo budget. Later that same year, NASA was forced to confirm that the test flight of the three-manned Apollo capsule planned for March 1965 was being rescheduled for late 1965 or early 1966. Most of the delay was attributed to "technological differences," a euphemism for coordinating the vast network of researchers, developers, and manufacturers. Finally, Kennedy went to the extreme of proposing a joint Soviet–US mission to the moon. The proposal was rejected by Congress. If Kennedy used the cooperation as a ploy, it succeeded; while rebuffing the President, Congress approved funding of NASA without reduction, allowing development to continue unabated.

Testing, Testing, and Even More Testing

James E. Webb, the NASA chief, put it best when he said that landing a US astronaut on the moon by 1970 was "a fast-paced program, not a crash program."[3] Webb probably didn't even realize his double entendre. True, the Soviets had the lead, but a consistent and error-minimizing process was the sure ticket to a lunar landing.

In order to minimize the risk of an unacceptable event occurring, the development of the Saturn V was defined by constant, rigorous, and detailed testing of each of the rocket's components. The mammoth size of the Saturn V meant that testing would occur on a greater scale than on any other project in human history. The comprehensive testing of components, sub-systems, and total systems was undertaken by both the twenty thousand contractors and NASA before the launch of any Saturn V. The guidelines from von Braun's office at the Marshall Space Center were clear, concise, and unwavering.

1. No test in the launch vehicle program should be eliminated for the sake of shortening the time schedule.

2. The "high risk concept" should be banished from our philosophy entirely in making time schedule in favor of a more conservative and realistic approach, which will finally lead to a more economic and successful program, and a shorter over-all time schedule in the end.

3. Especially for the early launchings of the big boosters, ample time should be allowed in order to arrive at a "mature" design with possible alternate solutions to be explored. "Mature" design meant one which is thoroughly investigated, well planned and based on proven design concepts with ample margins of safety.

4. No launching should be attempted unless great (though not absolute) confidence exists by all parties involved that it will be successful. To achieve this confidence, each and every component and sub-system must be carried through a Qualification Test Program, all systems tests must be made and evaluated, and painstaking inspection, including in-process inspection, must be performed.

5. Development tests such as wind tunnel investigations, structural tests, dynamic tests, static, battleship, all systems tests (long duration), etc. have to be performed according to a definite plan before first launching.

6. No stage should arrive at Cape Canaveral, as a rule, which is not complete and entirely checked out in the home plant to the satisfaction of the Government. Neglecting this rule will show earlier delivery date to the Cape, which many people favor because they assume this also means an early and successful launch.

7. Developing difficulties, including mating problems, should be straightened out in the home plant as far as possible and not at the Cape.

8. In programs like the manned Lunar Landing, the time schedule has to be ambitious—very ambitious. However, the time compression should not go beyond reality, lest it erode the spirit and the morale of the people involved: nothing is gained if the people at the working level believe that a schedule is unrealistic anyway, and therefore meaningless. It has been our long-standing experience that impossible deadlines and milestones are not taken seriously and only serve to undermine the sense of responsibility of the individual and his respect for those responsible for either making or accepting such schedules.

9. Time schedules that are established must be compatible with Government management capabilities, policies, procedures, controls, limitations, evaluations, re-evaluations, etc. A crash program or accelerated schedule makes sense only if it is in tune with the funding level appropriated by Congress. Moreover, accelerated programs also require certain relaxation in laws, regulations and practices in fields such as procurement, funding, facility planning, civil service personnel ceiling control, etc.[4]

Sara Caldwell was part of the massive testing regimen of the Saturn family of rockets. She represented the new wave of female aerospace engineers. She was one of only two women who worked on the first stage of the rocket, known as the S1-C.

A Louisiana girl, Sara attended Byrd High School in Shreveport. In March 1961, while a senior, she got the space bug. President Kennedy's speech was one turning point. "When I watched his speech, I said, 'I'm going to be part of that.'" A few months later, Sara and her best friend, A. W. Steed, were attending a reception for military brass at nearby Barksdale Air Force Base. Suddenly, Werner von Braun entered the room. Overcome with youthful exuberance, A. W. and Sara ran over to von Braun and started talking with him. Somehow the conversation turned to careers, and Sara told von Braun she wanted to be part of the space program, but there weren't many opportunities for young ladies. Von Braun would have none of it, telling the starry-eyed Sara, "Girls can do anything."[5]

Sara did just that. She studied mathematics at Louisiana State University and graduated in 1965. Her timing could not have been better because the Boeing Corporation, the prime contractor for the S1-C stage of the Saturn rocket, was hiring engineers at its facility in Michoud, Louisiana. Sara completed the twelve-page job application and underwent an extensive background check. She got a job as a systems test engineer. Sara was called upon to analyze data after the test firing of the S1-C stage, a firing that took take place at NASA's new Mississippi test facility.

In October 1961, the federal government selected an area in Hancock County, Mississippi, to be the site of a static test facility for launch vehicles for the Apollo manned lunar landing program. The construction project was the largest in the state of Mississippi and the second largest in the United States at the time. The selection of the site in Mississippi was a logical and practical one. The land was chosen because of its water access, essential for transporting large rocket stages, components, and loads of propellant. The area also provided a 13,500-acre test facility with a sound buffer of close to 125,000 acres.

After the S1-C firing, rolls of paper with scores of data points were brought to the Michoud facility and laid out on long tables where Sara analyzed the data, looking for any abnormalities. If Sara or one of her teammates discovered anomalies, he/she would first seek to identify the source of the problem and then make recommendations to resolve the difficulty. This routine lasted until NASA began manned Saturn V launches. Sara was still employed by Boeing under a federal contract, but she was now "...working for our astronauts, because we wanted them to be safe in their journey, ultimately to the moon and back."[6]

The Supply Chain Web

Anyone who has played the MBA simulation called the "Beer Game" understands that production and order quantities rarely reach a happy equilibrium resulting in over and under supply of a product. The Saturn program's dispersed research, production, and testing facilities, spread all over the country, presented a major logistics and quality-control obstacle. Imagine building pieces of a puzzle at various locations and then hoping all the pieces would fit together without a problem. Chances for success seemed dependent on too many variables.

The Saturn's first stage, as previously mentioned, was contracted to Boeing. The company was charged with testing and delivering a completed first stage to NASA in Florida, where the stage would be paired with the second and third stages before launch. The first stage was assembled outside of New Orleans, so every supplier (see the table below) had to be sure to coordinate the development, testing, and delivery of its components on a very tight schedule and under meticulous specifications. There were no personal computers. Fax machines didn't become popular until the late 1970s. Researchers couldn't scan images and attach them to e-mails. The modus operandi of the era was investigation with slide rules, pencils, and paper.

Boeing's Major Subcontractors

Subcontractor	Location	Components
Aeroquip Corp.	Jackson, MI	Couplings, pneumatic, and hydraulic hoses
Aircraft Products	Dallas, TX	Machined parts
AiResearch Manufacturing Co.	Phoenix, AZ	Valves

Boeing's Major Subcontractors

Subcontractor	Location	Components
Applied Dynamics, Inc.	Ann Arbor, MI	Analog computers
Arrowhead Products, Div. of Federal-Mogul Corp.	Los Alamitos, CA	Ducts
The Bendix Corp., Pioneer-Central Div.	Davenport, IA	Loading systems and cutoff sensors
Bourns, Inc., Instrument Div.	Riverside, CA	Pressure transducers
Brown Engineering Co., Inc.	Lewisburg, TN	Multiplexer equipment
The J. C. Carter Co.	Costa Mesa, CA	Solenoid valves
Consolidated Controls Corp.	Bethel, CT Los Angeles, CA	Pressure switches, transducers, and valves
The Eagle-Picher Co., Chemical and Metals Div.	Joplin, MO	Batteries
Electro Development Corp.	Seattle, WA	AC and DC amplifiers
Flexible Tubing Corp.	Anaheim, CA	Ducts
Flexonics, Div. of Calumet and Hecla, Inc.	Bartlett, IL	Ducts
General Precision, Inc., Link Ordnance Div.	Sunnyvale, CA	Propellant dispersion systems
Gulton Industries, C. G. Electronics Div.	Albuquerque, NM	Wiring boards
Hayes International Corp.	Birmingham, AL	Auxiliary nitrogen supply units
Hydraulic Research and Manufacturing Co.	Burbank, CA	Servoactuators and filter manifolds

Boeing's Major Subcontractors

Subcontractor	Location	Components
Johns-Manville Sales Corp.	Manville, NJ	Insulation
Kinetics Corporation of California	Solano Beach, CA	Power transfer switches
Ling-Temco-Vought, Inc.	Dallas, TX	Skins, emergency drains, and heat shield curtains
Marotta Valve Corp.	Boonton, NJ	Valves
Martin Marietta Corp.	Baltimore, MD	Helium bottles
Moog, Inc.	East Aurora, NY	Servoactuators
Navan Products, Inc.	El Segundo, CA	Seals
Parker Aircraft Co.	Los Angeles, CA	Valves
Parker Seal Co.	Culver City, CA	Seals
Parsons Corp.	Traverse City, MI	Tunnel assemblies
Precision Sheet Metal Inc.	Los Angeles, CA	Filter screens, anti-vortex, and adapter assemblies
Purolator Products, Inc., Western Div.	Newbury Park, CA	Umbilical couplings
Kandall Engineering Co.	Los Angeles, CA	Valves
Raytheon Co.	Waltham, MA	Cathode ray tube display system
Rohr Corp.	Chula Vista, CA	Heat shields

Boeing's Major Subcontractors

Subcontractor	Location	Components
Servonic Instruments, Inc.	Costa Mesa, CA	Pressure transducers
Solar, Div. of International Harvester	San Diego, CA	Ducts
Southwestern Industries, Inc.	Los Angeles, CA	Calips pressure switches
Space Craft, Inc.	Huntsville, AL	Converters
Stainless Steel Products, Inc.	Burbank, CA	Ducts
Standard Controls, Inc.	Seattle, WA	Pressure transducers
Statham Instruments, Inc.	Los Angeles, CA	Pressure transducers
Sterer Engineering and Manufacturing Co.	Los Angeles, CA	Valves
Stresskin Products Co.	Costa Mesa, CA	Insulation
Systron-Donner Corp.	Concord, CA	Servo accelerometers
Thiokol Chemical Corp., Elkton Div.	Elkton, MD	Retrorockets
Trans-Sonics, Inc.	Burlington, MA	Measuring systems and thermometers
Unidynamics/St. Louis, Div. of UMC Industries, Inc.	St. Louis, MO	Spools, harnesses, and ducts
United Control Corp.	Redmond, WA	Ordnance devices and control assemblies
Vacco Industries	South El Monte, CA	Filters, relief valves, and regulators

Boeing's Major Subcontractors

Subcontractor	Location	Components
Whittaker Corp.	Chatsworth, CA	Valves and gyros
Fred D. Wright Co., Inc.	Nashville, TN	Support assemblies and measuring racks

Source: Roger E. Bilstein, *Stages to Saturn A Technological History of the Apollo/Saturn Launch Vehicles* (Washington, DC: NASA History Office, 1996), pp. 424−26.

The coordination of the second and third stages, along with the Saturn V's computer instrumentation unit, presented similar planning and management challenges. Time was not always an ally. Perhaps there is a lesson in the Saturn project that transcends technology and engineering. Fear drives people to greatness as much as genius. Facing a future of Soviet scientific dominance, America was able to move with a purpose.

Pressure never seemed to let up. Pressure was related to the narrow window for each launch. A missed window meant a flight delay—the window couldn't be opened at will. Pressure was determined by the position of the launch coordinates (Complex 39 at Kennedy Space Center) and the selected moon landing site. Pressure ruled out launching during half of a given month because the landing site had to be in sunlight for the astronaut's descent. Pressure was inherent in the Lunar Excursion Module and the astronauts' suit design qualities that restricted available launch dates

Paul Thomarios's parents, Nickitas and Stilyani, in the 1950s. Paul learned a lot from his Greek immigrant parents, whose story was one of survival, sacrifice, and hard work. *Photo courtesy of Thomarios.*

Paul's third grade teacher, Jean Hixson, was one of the Mercury 13 and the second woman to break the sound barrier. Along with Paul's mother, Hixson was an inspirational figure in Thomarios's life. *Photo courtesy of Thomarios.*

During a November 16, 1963 visit to Cape Canaveral, President John F. Kennedy was briefed by famed engineer, Werner Von Braun, on the upcoming launch of a Saturn rocket. President Kennedy was assassinated in Dallas six days later. *Photo courtesy of NASA.*

The first stage of the Saturn V, without engines, was assembled at Boeing's Michoud, Louisiana, facility. The facility remains one of the larger manufacturing centers in the world. *Photo courtesy of NASA.*

Originally constructed to build Higgins Boats during World War II, the Michoud plant was taken over by Boeing Corporation during the Apollo program. *Photo courtesy of NASA.*

Supply chain innovation was a key component of the Saturn V's manufacturing process. A special barge was built to transport stages of the Saturn to the Kennedy Space center. *Photo courtesy of NASA.*

Technologically sophisticated for its time, the Saturn V's instrument unit, built by IBM, was a cutting-edge breakthrough. A mobile phone of the present era has exponentially more computing power than the "brains" of the Saturn V. *Photo courtesy of NASA.*

NASA developed a fleet of odd-looking aircraft, including the Pregnant Guppy, to transport the third stage of the Saturn V and the Apollo Service Module, Command Module, and Lunar Excursion Module (LEM) from California to Florida. *Photo courtesy of NASA.*

The F1 engines manufactured by Rocketdyne were, and still are, the largest and most powerful liquid-fueled engines ever built. Five F1s were used to launch the Saturn V. *Photo courtesy of Pratt Whitney Rocketdyne.*

The J2 engines powered the second and third stages of the Saturn V and were manufactured by Rocketdyne in Canoga Park, California. *Photo courtesy of Pratt Whitney Rocketdyne.*

Top Left: The first Saturn V, Apollo 4, goes through its final 36-story assembly in the massive Vehicle Assembly Building (VAB). *Photo courtesy of NASA.*

Top Right: One of the Saturn Vs being readied for rollout to the launch pad. Workers can be seen in the lower right hand corner. *Photo courtesy of NASA.*

Bottom Left: Apollo 11 lifting off on July 16, 1969. This mission landed the first human on the moon and fulfilled Kennedy's challenge. *Photo courtesy of NASA.*

Bottom Right: Apollo 11 after lift-off. The cloud between the first and second stages was caused by the liquid hydrogen fuel building up prior to separation. *Photo courtesy of NASA.*

Left: To prepare for the U.S. bicentennial, a Saturn V was moved out in front of the VAB. *Photo courtesy of NASA.*

Right: Apollo 12 on the launch pad on November 14, 1969. During lift-off, it was struck by lightning twice, and almost lost all power, but all systems successfully rebooted. *Photo courtesy of NASA.*

Bottom: In 1985, NASA displayed history's two most prominent launch vehicles, the Space Shuttle *Enterprise* and the Saturn V rocket, in the parking lot in front of the VAB. *Photo courtesy of NASA.*

Left: The Saturn rocket was dangerously close to being unsalvageable when the restoration of the rocket began in 1996. *Photo courtesy of Thomarios.*

Right: A portable air structure resembling an indoor tennis court was used to work on the second and third stages of the Saturn V. Here, workers prepare to set up the structure. *Photo courtesy of Thomarios.*

Bottom: The work on Stage 1 was performed outdoors. The rear section was separated from the rest of the first stage, due to the presence of asbestos in the heat shields. Scaffolding was built to allow access to the rocket. *Photo courtesy of Thomarios.*

Thomarios employee, Roy Campbell, posing with one of the wheels that came from the Stage 1 transport vehicle. *Photo courtesy of Thomarios.*

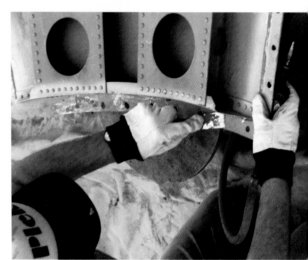

Removing more than twenty years of corrosion was a tedious, arduous task. A worker cleans an access panel on Stage 1. *Photo courtesy of Thomarios.*

A view of the asbestos-coated heat panels on the rear portion of Stage 1. *Photo courtesy of Thomarios.*

A worker removing asbestos from the steel-plated heat shields on the aft portion of Stage 1. *Photo courtesy of Thomarios.*

Mike Cioca installs replica, asbestos-free heat panels fabricated by Thomarios. *Photo courtesy of Thomarios.*

A view of the effects of twenty years of corrosion on one of the Saturn's F1 engines. *Photo courtesy of Thomarios.*

Thomarios employee and graphics expert, Gary "The Brush" Hagan, stencils the stars on the American flag on Stage 1. *Photo courtesy of Thomarios.*

An image of Stage 1 with all of the asbestos and paint removed. *Photo courtesy of Thomarios.*

Photos on right: A worker repairs holes in Stage 2. The punctures resulted from birds pecking through the stage's lightweight material. *Photo courtesy of Thomarios.*

Left: Paul Thomarios oversees the removal of black mold from Stage 2. Nick Bolea pressure washes the exterior of the rockets. *Photo courtesy of Thomarios.*

A Thomarios employee cleans corrosion from the inside of Stage 2. *Photo courtesy of Thomarios.*

Workers remove corrosion from Stage 2. *Photo courtesy of Thomarios.*

Left: A worker repairs corroded areas with an epoxy filler. *Photo courtesy of Thomarios.*

Right: After the corrosion was removed, employees thoroughly sanded the surface before priming and painting began. *Photo courtesy of Thomarios.*

Bottom: Rust and corrosion on the original bolts from the handling rails on Stage 2. *Photo courtesy of Thomarios.*

Left: A worker performs a test to ensure the adhesion quality of new paint on Stage 2. *Photo courtesy of Thomarios.*

Right: Nick Bolea inspects the inner ring of Stage 2. *Photo courtesy of Thomarios.*

Bottom: Nick Bolea paints Stage 2. *Photo courtesy of Thomarios.*

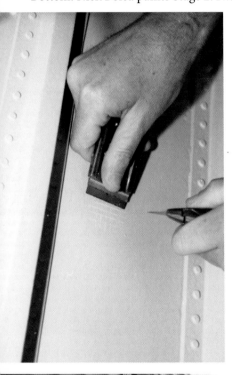

Top: Thomarios employee and fabricator Tom Casanova cleans the electrical connections on Stage 2. *Photo courtesy of Thomarios.*

Left: Russ Masters paints an area inside of Stage 3. *Photo courtesy of Thomarios.*

Right: Thomarios employees color-coat the wiring on Stage 2. *Photo courtesy of Thomarios.*

A view of Stage 3, which contained the Lunar Excursion Module (LEM). *Photo courtesy of Thomarios.*

A Thomarios employee cleans the Command Module and Escape Tower. *Photo courtesy of Thomarios.*

Top: A Thomarios employee removes an old rivet on the aft end of Stage 3. *Photo courtesy of Thomarios.*

Bottom: Paul Thomarios, foreground, speaks with Thomarios employees and Al Bachmeier, a representative from the Smithsonian Institution (blue helmet). *Photo courtesy of Thomarios.*

Left: An inspector monitors the dry film thickness of a coating on Stage 3. *Photo courtesy of Thomarios.*

Right: Beyel Brothers, Inc. employee maneuvering equipment that was used to move the stages of the Saturn V rocket from the parking lot to the Apollo/Saturn V Center. *Photo courtesy of Thomarios.*

Bottom: Preparing to move the enormous, refinished Stage 1 of the Saturn V. *Photo courtesy of Thomarios.*

Above: Stage 1 moves to the Apollo/Saturn V center. The short journey took nine hours. *Photo courtesy of Thomarios.*

Right: Stage 1 moves to the Apollo/Saturn V Center. *Photo courtesy of Thomarios.*

Stages 2, 3, and the Command Module are readied for transport to the Apollo/ Saturn V Center. *Photo courtesy of Thomarios.*

Left: Stage 2 of the Saturn V is readied for transport to the Apollo/Saturn V Center. *Photo courtesy of Thomarios.*

Right: The beginning of the move of stages 2 and 3. It took twenty hours to move the complete rocket to its new home 1.9 miles away. *Photo courtesy of Thomarios.*

Stages 2, 3, and the Service Module arrive at the Apollo/Saturn V Center. *Photo courtesy of Thomarios.*

For five months in early 1996, this parking lot was the workshop for Paul Thomarios and many of his employees. *Photo courtesy of Thomarios.*

A rendering of the Apollo/Saturn V Center at Kennedy Space Center, where the Saturn V is located. *Photo courtesy of Thomarios.*

Concept art for the pre-show area of the Apollo/Saturn V Center. This area greets guests, gathers them into a group, and presents a background story before ushering them into The Firing Room. © *BRC Imagination Arts, used by permission.*

Concept art for The Firing Room presentation in the Apollo Saturn V Center. This side view shows the angled windows above and behind the audience, which rattle during "launch". © *BRC Imagination Arts, used by permission.*

Concept art for Rocket Plaza and the display of the Saturn V rocket. © *BRC Imagination Arts, used by permission.*

Concept art for exhibits and displays to be placed alongside the Saturn V rocket in Rocket Plaza. © *BRC Imagination Arts, used by permission.*

Concept art for the un-built Return to Earth presentation. This show would have presented the Apollo 11 landing from the point of view of the Command Module. © *BRC Imagination Arts, used by permission.*

even further.[7] Pressure was Florida's unpredictable weather making the launch window even smaller.

The Saturn V Fleet

Given the small margin of error, the builders of the Saturn V were compelled to pay particular attention to the supply chain. Not only did NASA personnel have to manufacture a massive piece of technology, but they had to innovate the distribution and logistics process. In order to verify the space-readiness of the Saturn stages, NASA floated the boosters on massive barges from the Michoud plant to the Huntsville, Alabama, test firing site, back to Michoud for redesign, and finally off to Cape Canaveral in Florida. The cost of the entire trip was almost $10 million in today's dollars. One of the barges that carried the Saturn booster was called the *Palaemon*, for the Greek god who was the protector of ships. The barge was 180 feet long, 38 feet wide, and was as tall as a four-story building.

The *Palaemon* was part of the NASA water transport fleet, which included three 270,000 gallon liquid hydrogen barges and six 105,000 gallon liquid oxygen barges (the fuels used by the Saturn V engines), two barges for transporting the Saturn V engines on inland waterways, called the *Little Lake* and the *Pearl River*, a seagoing vessel for transporting rockets from California, called the *Point Barrow*, and a number of towboats, including the custom-built *Clermont*, the

Apollo, and the *Bob Fuqua*. Before the Saturn rockets voyaged into space, their components had already logged thousands of miles over the Pacific and the Atlantic Oceans and required negotiation of the Panama Canal, the Gulf of Mexico, and the Intercoastal Waterway. The waterborne routes were time-consuming, but remained the only feasible mode of transporting the larger of the Saturn stages.

Another transportation breakthrough came in the form of the Pregnant Guppy.[8] The Pregnant Guppy was an outsized cargo aircraft that incorporated the wings, engines, lower fuselage, and tail from a Boeing 377 Stratocruiser, with a huge upper fuselage more than 20 feet in diameter. The aircraft had a volume of 22,500 cubic feet and was built to transport outsized cargo for NASA's Apollo program. In the early 1960s, the plane airlifted an S-IV stage of the Saturn I at a weight of almost 21,000 pounds, which at that time was the largest cargo ever transported by air. The Guppy added to NASA's transport capabilities and was critical in shrinking the delivery time of Saturn components.

The space program compelled design innovation across the manufacturing board. Necessity drove much of the novel advances. While the aim of the program was to get a man to the moon, life on Earth shared in the technology improvements. The blow rubber molding process reshaped how athletic shoes were manufactured. The need for cordless tools on the moon spawned a new industry. Telemetering devices were used to improve the medical diagnostic process. Computers used to manage the flight planning process became key in large-scale building construction mapping.

In 1960, Wernher von Braun said he was "determined to fly [him]self with one of the first space ships to the moon."[9] Obviously he didn't, but it was this unbridled bravado that colored NASA. The forward-looking enterprise became a great achievement because, as James Webb said, the race to the moon "...was going to be a national project [and] it had to engage the best people in the nation."[10] Sure, there were periods of disappointment and worry. Thomas O. Paine, a NASA administrator was quick to point out that when "...we first sent, in 1968, Frank Borman...around the moon, some people brought up the rather chilling feeling that if we really goofed and he got stuck in lunar orbit, every time a pair of lovers saw the moon at night, they'd have to think of those astronauts going around it."[11] Then again, Julian Scheer, NASA's assistant administrator for public affairs from 1962 to 1971, understood the program's breadth. "Think what it took to accomplish America's seven manned landings on the moon: an industrial infrastructure of twenty-five thousand companies, many of which had not previously existed; technical, engineering and scientific programs on college campuses and in private labs; breakthroughs in propulsion, in tracking and data acquisition, in human engineering and medicine."[12] Or maybe the whole project came down to the nature of human evolution as William Pickering, the head of NASA's Jet Propulsion Laboratory from 1954-1976 noted, "The difference between the stone age and now is curiosity, and curiosity has always been the driving force to advancement."[13]

1. "U.S. Rocket Wrecked," *Chicago Tribune*, December 7, 1957.
2. Dwayne A. Day, "Saturn's fury: effects of a Saturn 5 launch pad explosion," *The Space Review*, accessed March 1, 2010, http://www.thespacereview.com/article/591/1.
3. Piers Bizony, *The Man Who Ran the Moon* (New York: Thunder's Mouth Press, 2006), p. 87.
4. H. H. Koelle, *Learning from the Past: Birth, Life and Death of the Saturn Launch Vehicles* (Berlin: Technical University Berlin, Institute of Aeronautics and Astronautics, 2001), p. 27–28.
5. Sara Caldwell, interview by Andrew R. Thomas, January 15, 2010.
6. Ibid.
7. Robin Wheeler, "Apollo lunar landing launch window: The controlling factors and constraints," *Apollo Flight Journal*, 2009, http://history.nasa.gov/afj/launchwindow/lw1.html, accessed March 1, 2010.
8. Roger E. Bilstein, *Stages to Saturn: A Technological History of the Apollo/Saturn Launch Vehicles* (Washington, DC: NASA History Office, 1996), p. 319.
9. Wernher von Braun, *First Men to the Moon* (New York: Holt, Rinehart, and Winston, 1960), p. 4.
10. Bizony, *The Man Who Ran the Moon*, p. 92.
11. Glen Swanson, *Before This Decade is Out* (Gainesville: University Press of Florida, 2002), p. 16.
12. Ibid., p. 74.
13. Ibid., p. 86.

My mother, Stilyani Kostandakis, was born on January 5, 1914, on the island of Koutalis, in the Sea of Marmara. Early in 1922, along with her mother, father, and three brothers, she, like my father, became a refugee in Lavrion, Greece. Soon after the family's arrival in Greece, her father, Apostolos, passed away. My grandfather died of a broken heart after being uprooted.

My Story

Paul N. Thomarios

On August 15, 1937, my parents married. Both of my grandmothers were widowed. When my father departed for his three-year stint in the Merchant Marines, he left knowing his wife was pregnant. On May 28, 1938, my brother Gregory was born.

When Greece fell in April 1941, everyone was forced into labor for the Nazis. My mother lost all contact with my father. She had no idea where he was or what he was doing. My father tried to get letters through, but to no avail—incoming mail was highly censored by the Nazis.

When the war ended in 1945, an army buddy of my father, Victor White, wrote a simple postcard in English and mailed it to my mother. The information was exactly to the point. "Wife, I am in the US Army; will contact you soon." Upon receiving the postcard, Aristidas Kostandakis, my mother's brother and my uncle, walked almost forty miles to the British Embassy in Athens, because he was told the postcard was written in "English." The British figured out that the sender was an American and let my uncle know that until an American consulate was set up, not much could be done. My uncle had to walk back and deliver this news to my mother.

An American GI later drove into Lavrion, in search of Stilyani and her son. My father's wife and his never-before-seen son were soon classified by the US Military as "American Dependents;" my father had become an American citizen during the war. My mother and brother were flown from Athens to Naples, where Stilyani was misclassified as an Italian war bride, but nonetheless they gained passage on a Liberty ship, the *F. Thomas Berry*, and sailed for New York. As destiny would have it, F. Thomas Berry was the captain of the airship *Akron*. Stilyani and Greg arrived in New York on June 20, 1946. They headed off to Akron, where Stilyani reunited with her husband and Greg met his father for the first time. A new life began.

Chapter 5
Magnificent Deterioration

The sudden disappointment of a hope leaves a scar
which the ultimate fulfillment of that hope never
entirely removes.
—Thomas Hardy

Michael Collins, who flew the command module while Neil Armstrong and Buzz Aldrin walked on the moon's surface, was born in Rome, Italy. Collins was an Air Force brat and later joined up to become a pilot. His lonely lunar orbit on July 20, 1969, was probably quite comfortable, since living in so many places left him without a hometown. The Apollo mission was just another stop along his life's trajectory.

Back on Earth, the Vietnam War was still being fought, but earlier in July, the first US troops were withdrawn from the conflict. Ted Kennedy was answering questions about what had happened at Chappaquiddick. Reports were being released that the nation was facing a shortage of fifty thousand computer programmers. Disney's *Love Bug* was in theaters. A retailer was running a special on eight-track tapes for $5.88 each. Color television set makers were telling customers to buy their latest models to get the best view of Neil Armstrong's first step onto the lunar surface. Collectors could get a set of Apollo coins adorned with a headshot of the astronauts. For some unknown reason, the coins were being newly minted at the Bavarian State Mint in Munich, Germany.

Today, Baby Boomers still rate the moon landing as the most significant event in the last fifty years. Younger generations aren't as sure, ranking the 2008 election of Barack Obama equally as significant. In 1999, the Gallup organization asked Americans to identify the most important events of the century. The moon landing ranked seventh. World War II was first. No doubt, though, Neil Armstrong's name will be a part of the planet's vocabulary, like Thomas Edison, Albert Einstein, and the Wright brothers, names that carry meaning beyond simple pronunciation.

The real history of a people, and in fact all humankind, is not in prices and wages, nor in elections and battles, nor even in the tenor of the common man; it is the long-lasting contributions made by great people to the sum of human civilization and culture.[1] Undeniably, the great events were and are built like the pyramids, at times on the backs of thousands of common men. Still, the history of a country like England is the record of her exceptional men and women, her inventors, scientists, statesmen, poets, artists, musicians, and philosophers; of the additions they made to the technology, wisdom, artistry, and decency of their people and all of humankind.[2]

For future generations, the history of the US space program will be linked permanently to Neil Armstrong, like other histories are connected to Johannes Gutenberg, Henry Ford, and Bill Gates. Armstrong's first step has been rebroadcast so many times that it has become the iconic image of man's great adventure into outer space. Until another extraterrestrial event overtakes the imagination of the global populace, the giant leap will have primacy.

From the Earth to the Moon and to the Back Lot

A few days after Apollo 11 successfully returned, NASA issued a tentative planning schedule for the next nine Apollo missions.[3]

Flight	Launch Plans	Tentative Landing Area
Apollo 12	November 1969	Oceanus Procellarum lunar lowlands
Apollo 13	March 1970	Fra Mauro highlands

Flight	Launch Plans	Tentative Landing Area
Apollo 14	July 1970	Crater Censorinus highlands
Apollo 15	November 1970	Littrow volcanic area
Apollo 16	April 1971	Crater Tycho (Surveyor VII impact area)
Apollo 17	September 1971	Marius Hills volcanic domes
Apollo 18	February 1972	Schröter's Valley, river-like channel ways
Apollo 19	July 1972	Hyginus Rille region, Linear Rille, crater area
Apollo 20	December 1972	Crater Copernicus, large crater impact area

Vice President Spiro Agnew proposed landing men on Mars by the end of the century. However, with Kennedy's mission fulfilled, questions were asked about the future of the Apollo program. The US economic system faced the highest peacetime inflation in its history. NASA felt the heat in the early 1970s, with a budget cut of 12.5 percent, creating a loss of 175,000 jobs across the aerospace industry. The Apollo program was the hardest hit.

Apollo 20 was cancelled in January 1970. In the same month, NASA Administrator Thomas Paine announced that Saturn V manufacturing would halt with vehicle 515, the rocket that was to be used for Apollo 20, but eventually was converted into Skylab Orbital Workshop backup hardware (now on display at the National Air and Space Museum).[4] The flights planned for Apollo 15 and Apollo 19 were scrubbed in September 1970, and the remaining missions were renumbered 15 through 17. In late October 1970, reports surfaced that Boeing had overcharged for the Saturn boosters it had produced. It appeared the space program had reached the final frontier.

In total, fifteen Saturn Vs were built and twelve were launched. Two were unmanned: Apollo 4 (the first launch of a Saturn V) and Apollo 6. Ten carried astronauts: Apollo 8, 9, and 10 were designed to test the Saturn V and its spacecraft before the "moon shot." Six missions landed Americans on the moon: Apollo 11, 12, 14, 15, 16, and 17. One mission, Apollo 13, failed to reach the moon, but did reach theaters decades later, in a film starring Tom Hanks. By 1972, NASA's manned lunar launches were over.

NASA relegated each of the remaining Saturn V rockets to the back lots of the Kennedy Space Center in Florida, Johnson Space Center in Houston, and Marshall Space Flight Center in Huntsville, like stars past their prime. In 1969, the Marshall Center put its Saturn V outside for display. Kennedy Space Center dismantled the stages of Apollo 18 in 1972, and reassembled them in front of the massive Vehicle Assembly Building in 1975. In 1977, the Johnson Space Center put their Saturn V on display.

Built to withstand powerful launches, the rockets did not fare well outside in the heat and humidity.

They rotted—paint peeled off, parts rusted, and excess moisture led to structural failures. In early 1996, a team from the Smithsonian's National Air and Space Museum in Washington, DC, visited Kennedy Space Center to assess the condition of the deteriorating tourist attraction. Much of the rocket's exterior was made of aluminum, and aluminum corrodes quickly in air filled with salt. One of the Smithsonian inspectors, a chemist, explained, "When aluminum corrodes, it exfoliates into brittle layers that easily flake off, like pages in a decaying old book."[5] The space program wasn't a science fiction tale anymore. It had become a horror story. The Saturn rocket was pocked with gaping tears, rusted rivets, frayed wire, and fungi and other plant growths. The rocket was also littered with innumerable red berries and small fish bones brought in by blackbirds that had nested in the rocket.[6] The rocket's shiny exterior had become a mildewed green.

How Did We Arrive at This?

Much has been written about the abrupt end of the Apollo program. On the surface, America's long-term plans for the moon were cut short by the smaller budgets imposed by Congress and the Nixon administration after Apollo 11. NASA's spending had peaked in the mid-1960s. By January 1970, the original 400,000-strong workforce had shrunk to 190,000, and plans were in place to eliminate another 50,000 jobs almost immediately.[7] According to Roger Launius, senior curator at the Smithsonian National Air and Space Museum, the growing lack of money wasn't the only cause. NASA officials weren't sure the remaining money was being well spent.[8]

Safety considerations were also a growing concern. In April 1970, the mechanical failures of the service module that threatened the lives of the crew of Apollo 13 heightened the risks involved in continued lunar missions. Many within NASA believed a catastrophe on the launch pad or during a mission was almost a certainty.[9]

The availability of funding led to some difficult decisions. One consequence of curbing the Apollo program was to redeploy a heavy-lift Saturn V to launch the Skylab orbital station in 1973. The prospective development of a space shuttle—endorsed by President Nixon in 1969—had begun diverting attention at the agency, as well.[10]

While it is convenient to point fingers at Congress, President Nixon, or some unnamed NASA bureaucrat for the inglorious end to America's moon program, the real answer has deeper roots. As with many events in life, an ending is related to a beginning. President Kennedy's impact on the space program had far-reaching effects.

Kennedy's challenge in May 1961 to send men to the moon was contrary to his earlier statements and actions. In his inaugural address that January, the President spoke directly to the Soviets, saying "together let us explore the stars."[11] In his State of the Union message ten days later, the President invited the Soviet Union "to join with us in developing a weather prediction program, in a new communications satellite program, and in preparation for probing the distant planets of Mars and Venus, probes which may someday unlock the deepest secrets of the universe."[12] On January 31, under orders of the President, the chief of the US Weather Bureau sent a cable to his Soviet counterpart, inviting him to a World Meteorological Organization meeting in Washington, DC, to discuss the uses of satellites for weather prediction.[13] The Soviet delegation, without explanation, declined to attend.[14]

In March, the President received a briefing from NASA Director James Webb about the direction of America's space program. In the document, Webb outlined the need for greater funding to fully develop the Saturn family of rockets, especially in light of the Soviet Union's space successes.

> The first priority of this country's space effort should be to improve as rapidly as possible our capability for boosting large spacecraft into orbit, since this is our greatest deficiency...We cannot regain the prestige we have lost without improving our present inferior booster capability, and doing it before the Russians make a major breakthrough into the multi-million pound thrust range.[15]

At that time, Webb did not indicate that landing an American on the moon was part of NASA's strategy. Endorsing Kennedy's view on space being a cooperative effort, Webb concluded:

> The extent to which we are leaders in space science and technology will, in some large measure, determine the extent to which we, as a nation, will be in a position to develop this emerging world force and make it the basis for new concepts and applications for nations willing to work with us in the years ahead.[16]

A few days later, the President approved an additional $56 million for Saturn vehicle development, which brought NASA's budget for the upcoming fiscal year to a conservative $1.235 billion.[17] The fledgling Project Apollo, which was introduced into NASA's budget during the Eisenhower years, did not receive any funding.[18] After the submission of the NASA budget to Congress on March 28, the *Washington Star* observed "If the United States is to get a 'new look' in space, it will have to wait at least another year for the change to begin."[19] Several events altered the time frame.

On April 12, the Soviet Union sent the first human being into space: Yuri Gagarin. The event itself was not unforeseen, as most of the world concluded that the Soviets were far ahead of the Americans in the space race. Reality, though, has a way of crystallizing public opinion.

Across America, criticism of the President's space policy was stark. The *New York Times* reflected the mood: "It is high time to discard this policy [the current space policy]. In fact, if the United States is to compete in space, we must decide to do so on a top-priority basis immediately, or we face a bleak future of more Soviet triumphs.... Only Presidential emphasis and direction will chart an American pathway to the stars."[20] On Capitol Hill, in an atmosphere of panic, nearing hysteria, hearings were ordered.[21] Republican James Fulton told NASA chief Webb, "I believe we are in a race...tell me how much money you need and this committee will authorize all you need."[22] Democratic Congressman Victor Anfuso announced, "To properly alert our people I am ready to see our country mobilized to a war-time basis because we are at war. I want to see our schedule cut in half. I want to see what NASA says it is going to do in ten years done in five. I want to see some firsts coming out of NASA, such as the landing on the moon."[23] Representative James G. Fulton of Pennsylvania put it more bluntly saying he was "Damned well tired of coming in second."[24]

Fear was a driving motive as well. Since the Soviets were capable of orbiting a man around the Earth with precision, they could alter the balance of power of the Cold War. What was preventing the Russians from putting atomic weapons into space? Senator

John Stennis of Mississippi concluded that if the Soviets could put one weapon into Earth orbit they could do so "with a whole lot of weapons simultaneously."[25]

Many Americans didn't know what to believe. One Wisconsin auto salesman thought the whole incident could be a propaganda ploy, but nonetheless his patriotism shone through. "I don't believe we are second best."[26]

The rest of the world was impressed with Gagarin's flight. One British scientist called the flight the greatest scientific advancement in the history of man. The French press "relegated all other news to a secondary position." The Vatican called the flight "a universal good," and a Swiss newspaper described the voyage as "the number one event of the Twentieth Century."[27]

A shrewd politician, Kennedy quickly began to see that his position on joint space exploration was a tenuous one at best. On April 14, just three days after Gagarin's flight and Kennedy's telegram to Khrushchev, Kennedy held a meeting with his top NASA officials and key aides to discuss the new threat of Soviet space supremacy. The President asked, "Is there any place we can catch up with them? What can we do? Can we put a man on the moon before them? When will Saturn be ready? Can we leapfrog?"[28]

NASA Deputy Director Dryden laid it out: "The one hope is a crash program similar to the Manhattan Project. But such an effort might cost $40 billion [about $300 billion today], and even so there was only an even chance of beating the Soviets."[29] Kennedy was shocked by the cost. This was a time when $40 billion was a lot of money. America's population was 40 percent less in 1961 than today; and her national wealth was about 80 percent less than the current GDP. Kennedy concluded the meeting by telling his staff to get more information, "When we know more, I can decide if it's worth it or not. There's nothing more important."[30]

As if the Gagarin mission wasn't enough, the United States suffered another severe blow to its world reputation in mid-April 1961. The "Bay of Pigs" operation, aimed at overthrowing the Castro regime, was launched on April 15. Air strikes ahead of the invasion force proved to be ineffective at hitting targets, leaving the Cuban Air Force intact. The April 17 landing force was strafed, pinned down, and faced twenty thousand mobilized Cuban troops. Kennedy pulled the plug on the mission and twelve hundred members of the invading force were captured, one hundred were killed, and the world was stunned by the entire incident. It was ill-timed, ill-conceived, and perhaps illegal.

Kennedy aide Ted Sorensen described the scene at the White House in the aftermath of the Bay of Pigs. The President was a depressed and lonely man "who knew [that] he had handed his critics a stick with which they would forever beat him; that his quick strides toward gaining the confidence of other nations had been set back; that Castro's shouting boasts would dangerously increase the Cold War frustrations of the American people."[31] The technology of sending a man to the moon was put into doubt because the United States had made a debacle of the invasion of Cuba.

A fellow New Englander would ultimately provide the comeback spark Kennedy needed. A loner even within the original seven astronaut community, Alan Shepard was born in East Derry, New Hampshire, a place a 1902 guidebook called the perfect spot for summer visitors because the town was quiet, clean, and neat. Shepard had the same traits and the gritty spirit to withstand the pressures and dangers of rocket flight. "As far as jeopardizing one's life goes, as soon as you get more than 50 or 60 feet off the ground, you're in trouble," he said.[32] Fortunately for NASA, Shepard's flight on May 5 was a success. The nation was relived and jubilant. The New Hampshire state legislature almost immediately designated East Derry as Spacetown, U.S.A. The turn of events reinspired Kennedy and he confirmed his intent to accelerate the country's space program.

> For while we cannot guarantee that we shall one day be first, we can guarantee that any failure to make this effort will make us last. We take an additional risk by making it in full view of the world, but as shown by the feat of astronaut Shepard, this

very risk enhances our stature when we are successful. But this is not merely a race. Space is open to
us now; and our eagerness to share its meaning is
not governed by the efforts of others. We go into
space because whatever mankind must undertake,
free men must fully share.[33]

The Apollo Paradox

The President's announcement sent shock waves
through NASA, which, during its short history, had
been a relatively small research and development agency.
However, Kennedy was not interested in going ahead
half way. Congress affirmed the President's plans with
few reservations. In 1960, NASA employed about ten
thousand people and NASA funding was 37 percent
of the entire space program's budget. By 1966, NASA
employed almost four hundred thousand people,
accounted for 76 percent of the space budget, and con-
sumed more than 5 percent of the entire US economy.[34]
The construction of the Panama Canal, a peacetime
project, and the Manhattan Project to build the atomic
bomb during World War II were the only projects
comparable to the scope and magnitude of the Apollo
program.[35]

If the growth of the Apollo program was hyper-
bolic, its demise was just as abrupt. America invest-
ed in the program and the outcome. The expansion
was unbridled, and like a business that expands too
quickly, when funding dried up, the program was

faced with too many facilities, too much equipment, and too many employees. The product of the program, flying men to the moon, was something the United States was not buying anymore. NASA needed to downsize and change its mission.

Of the seven original Mercury astronauts, Carpenter, Cooper, Glenn, Grissom, Schirra, Shepard, and Slayton, two are still living—Carpenter and Glenn. Shepard, the first American in space, also was the oldest American to set foot on the moon. At age seventy-seven in 1998, Glenn became the oldest American to fly in space aboard the space shuttle *Discovery*. Glenn saw the opportunities clearly during the 1959 press conference to introduce the Mercury astronauts; the space program was a reenactment of the Wright Brothers story on a different scale, with "Wilbur and Orville pitching a coin to see who was going to shove the other one off the hill down there. I think we stand on the verge of something as big and as expansive as that was fifty years ago."[36] Unfortunately, after the thrill diminished, the country was littered with remnants of the Apollo program, remnants that were not needed and became part of the space program's junkyard.

---

1. Will Durant, *The Greatest Minds and Ideas of All Time* (New York: Simon & Schuster, 2002), p. 6.

2. Ibid., p. 7.

3. Office of Manned Space Flight, NASA, "Manned Space Flight Weekly Report-July 28, 1969," NASA Historical Collection,

NASA Headquarters, Washington, DC.

4. Thomas O. Paine, "NASA Future Plans," NASA press conference transcript, January 13, 1970, NASA Historical Collection, NASA Headquarters, Washington, DC.

5. Frank Winter and Scott Wirz, "Saturn Rising," *Air and Space Smithsonian,* January 1997, p. 30.

6. Ibid.

7. Kenneth Silber, "Down to Earth: the Apollo Missions That Never Were," *Scientific American,* July 16, 2009.

8. Roger Launius, interview by Andrew R. Thomas, March 7, 2010.

9. Ibid.

10. Ibid.

11. S. Document 88-18, at 189 (1963).

12. Ibid.

13. John M. Logsdon, *The Decision to Go to the Moon: Project Apollo and the National Interest* (Cambridge, Massachusetts, MIT Press; 1970), p. 94.

14. Ibid.

15. James Webb, "Administrative Presentation to the President," March 21, 1961, NASA Historical Collection, NASA Headquarters, Washington, DC.

16. Ibid.

17. Logsdon, p. 99.

18. Ibid., p. 100.

19. *Washington Star,* March 29, 1961, p. 4.

20. *New York Times,* April 16, 1961, Section IV, p. 3.

21. Logsdon, p. 103.

22. *Discussion of Soviet Man-In-Space Shot: Hearings Before the United States H. Comm. On Science and Astronautics,* 87th Cong. 7 (1961).

23. Ibid., p. 13.

24. *The Bee* (Danville, VA), August 15, 1962, p. 5A.

25. John M. Logsdon, *The Decision to Go to the Moon: Project Apollo and the National Interest* (Cambridge, MA: MIT Press, 1970), p. 98.

26. Ibid.

27. *New York Times*, April 13, 1961, pp. 14–16.

28. Logsdon, p. 106.

29. Ibid.

30. Ibid.

31. Ibid., p. 111.

32. Loudon Wainwright, "The Old Man Gets His Shot at the Moon," *Life Magazine*, July 31, 1970, p. 56.

33. Logsdon, *The Decision to Go to the Moon*, p. 100.

34. Roger Launius, "The Apollo Decision as a Model of Public Policy Formation," *Air Power History*, Winter 2009, p. 22.

35. Ibid., p. 23.

36. Joseph N. Bell, "I lived with the astronauts," *Popular Mechanics*, December 1959, p. 262.

I was born on March 27, 1947, and named after my grandfather. My first home was a one-bedroom apartment at the corner of Arlington and Market Streets in Akron, Ohio. At the age of nine, I went to work for my father at

My Story

Paul N. Thomarios

his painting business. Greeks always seem to be good painters. In 1968, I enrolled at the University of Akron and graduated in 1972, with a degree in Political Science. Throughout college, I did a lot of house painting.

After graduation, I explored other careers, but none seemed to fit. I decided to stay in Akron and work in the family business. Over time, my father slowed down, and I began to run the company. By the mid-1970s, our customer base included our stalwart residential customers, but had also expanded to include industrial and commercial customers.

I was willing to try new things. If the opportunity arose, I'd hang wallpaper. I'd paint railroad and highway bridges. I'd do tank liners. I eventually became known as the painting contractor who would take on any challenge and do the job well. I guess I just couldn't say "no."

Chapter 6

Resurrection Begins

Nations, like stars, are entitled to eclipse. All is well, provided the light returns and the eclipse does not become endless night. Dawn and resurrection are synonymous. The reappearance of the light is the same as the survival of the soul.

—Victor Hugo

In Greek mythology, Apollo was originally regarded as a god of light. He withdrew in the winter, but came back in the spring to gladden the land with his brightness; he calmed the wintry sea after the equinoctial gales; he protected the ripening crops, and received in autumn the first-fruits of the harvest; he was the patron of flocks and pastures, and slew

rapacious beasts that came to do them harm. The
Apollo space program, and its moon landings, couldn't
do all things the Greek god did. The program, though,
was a very magnificent achievement in American and
world history.

On December 14, 1972, Apollo 17 Commander
Gene Cernan reflected on the end of an era. As he
took his last step from the moon, he wanted to put
the program into perspective. "I'd like to just (say)
what I believe history will record. That America's
challenge of today has forged man's destiny of tomor-
row." The final moon flight left the Russian writer,
Yevgeny Yevtushenko, disappointed. He wouldn't get
his wish of having a poet travel to the moon. Perhaps,
the Apollo program would have been better off with
a book of poetry, as well as continued funding. Neither
happened.

Cernan's family most likely had been to the Kennedy
Space Center Visitors Complex, which was originally built in 1967 so that NASA employees and their
families could view space center operations. Today,
the center draws 1.5 million visitors a year, of which
40 percent are from outside the United States. After
Apollo 17, NASA planned new missions including
the space station and shuttle projects. The remaining
Saturn V rockets took journeys of their own that were
far from intergalactic. One of the rockets would make
its way from near rubble to the Visitor's Center. The

journey was, in actuality, measured in feet not miles, but took decades to complete.

The first step came in anticipation of America's Bicentennial in 1976. The unassembled Saturn V at Kennedy Space Center was placed in a horizontal position in front of the Vehicle Assembly Building (VAB). In celebration of the bicentennial, The VAB and the Launch Control Center were opened to the public for the first time and the fifty-two story VAB was painted with the largest-ever American flag. Visitors to the Kennedy Space Center usually glimpsed the Saturn V while whizzing by on a tour bus. The rocket didn't leave much of an impression.

The Smithsonian Institution had an agreement with NASA to take the agency's artifacts. In NASA's case, artifacts were unique objects that documented the history of the science and technology of aeronautics and astronautics. Their significance and interest stemmed mainly from their relation to historic flights, programs, activities, or incidents; achievements or improvements in technology; our understanding of the universe; and important or well-known person-alities. In January 1979, under the NASA-National Air and Space Museum Artifacts Agreement, the Saturn V was turned over to the Smithsonian. Still, the Saturn V remained in the same place outside the VAB for the next sixteen years, and its decay was painted over and puttied up. The Smithsonian's mission was, in the words of senior curator, David DeVorkian, to make "the marvels of our day accessible to the citizens of the 25th century."[1] The Saturn was in danger of not being around for the twenty-first century.

The Giant Leap for the Saturn Rocket

In the early 1990s, NASA and TW Recreational Services (TWRS), the private firm that operated the concessions at the Kennedy Space Center, began planning to take the Saturn V in out of the rain. In 1994, TWRS invited architecture and engineering firms across America to submit proposals for building a protective cover for the Saturn V. The hopes were to provide a haven for the Saturn to preclude further decay.

The bidders provided more than were asked of them. Concepts arrived from world famous architects like Frank Gehry, who designed the Walt Disney Concert Hall in Los Angeles and I. M. Pei, the architect of Rock and Roll Hall of Fame in Cleveland, Ohio. These "starchitects" wanted to create sculptural masterpieces celebrating the greatness of the Saturn V.

The winning bidder, Morris Architects of Orlando, brought in BRC Imagination Arts of Burbank, California, because of their prior work with museums, NASA, and the Walt Disney Company. BRC developed and produced the story and media master plan, telling how the rocket would be featured at Kennedy Space

Center. In addition, a concept architect, Clive Grout, was added to create the look and feel of the building. Grout's style was summed up in a comment about the externalities impacting structure. "[E]ntertainment and experience will begin even more strongly to shape ways that architecture, design, merchandising, shopping, travel and leisure define themselves."[2] Morris would do all the building engineering and construction drawings.

In 1981, Bob Rogers founded BRC (Bob Roger's Company) in his garage and quickly became a global leader in the field of experience design and production. One of BRC's architectural team was clear about the impact of multi-media on the individual. "As a result of shorter attention spans, the 21st Century will need even better storytellers in cultural attractions. We must capture the public's imagination in less time and hold it for longer."[3] Prior to working on the Florida complex, BRC had designed the Space Center Houston, a breakthrough for NASA. The Houston site discarded the old practice of displaying artifacts with little context for or explanation of them. "The exploration of space is one of the greatest adventures of all time," said Rogers, paraphrasing President Kennedy. "We wanted to put the adventure into it." As one of the advertisements for the center indicated, "On your next trip, don't just leave town. Leave the planet!"[4]

BRC's proposal to NASA for the design and production of the Apollo/Saturn V Center at Kennedy
Space Center was quite pointed.

> If all you want to do is keep the rain off your rocket,
> stop reading. Select someone else. But if instead you
> want to take the next generation to the moon... If
> you want them to strap themselves on top of an
> unproven rocket containing more explosive power
> than an atom bomb... If you want them to feel the
> roar of the mighty Saturn V as it lifts off the pad... If
> you want them to feel the exhilarating thrill and
> awe of the greatest adventure of all time... If you
> want them to stand on the moon, shoulder to shoul-
> der with the astronauts and shed a tear as they look
> back at our home planet... Then keep reading. We
> can help.[5]

The project would be like trying to find water in a
desert with a divining rod. Apollo program parts
were strewn all over the Kennedy Space Center. The
launch umbilical tower's top half had been cut off and
lay rusting in a scrap yard. Under launch pad 39A,
to which the Teflon-coated Apollo's escape chute still
ran, piles of dust covered the area and a mouse had
built its nest out of discarded cigarette butts in the
mechanical oxygen generator. The remaining consoles
from the Firing Room and the command module
hadn't aged well and were housed in a distant, dark
building. The customized van that carried the astro-

nauts to the pad and the Lunar Lander was in another place. The Saturn V was laid out next to the giant Vehicle Assembly Building. To make matters worse, a rumor circulated that NASA had lost the blueprints for the Saturn rocket. The rumor was only partially true. Paper copies weren't available, but luckily the information was found on microfiche.

Creating a dramatic experience was a vital part of the BRC plan for Kennedy Space Center. One issue as Rogers noted, was the current configuration that made "the world's largest rocket seem small" because it was "right next to one of the world's largest buildings [VAB] and... people first glimpse[d] it from almost a mile away."[6] Instead, visitors needed to feel the texture and scope of a real launch. On arrival, visitors would be taken off the bus and into a re-creation of the original Firing Room. They would experience a dramatic account of the desperate race to the moon, including the risks and heroism involved. The emotional lift would be heightened with an exit from the Firing Room and arrival directly under the five huge rocket engines of the Saturn V. From there, the visitors could walk the entire 363-foot-length of massive launch vehicle.

In February 1995, ground was broken on the Apollo/Saturn V Center. Two months later, Delaware North Companies (DNC) assumed operation of the Kennedy Space Center Visitor Complex. Brothers Marvin,

Charles, and Louis Jacobs started DNC in 1915 by vending popcorn and peanuts in Buffalo, New York. In the late 1920s, the company landed the concession operations at Detroit's Navin Field. After the baseball season ended, the story goes, Louis Jacobs paid an additional $12,500 to the Detroit Tigers owner because the concessions brought in much more revenue than expected. Word spread of Jacobs' graciousness, and other baseball owners signed the company, now dubbed Sportservice, to handle concessions at their parks.

Current Delaware North Chairman and Chief Executive Officer Jeremy Jacobs, the son of founder Louis Jacobs, represents the second generation of the Jacobs family to own and lead the company. His three sons—Jerry Jr., Lou, and Charlie—hold executive positions with Delaware North. Under their guidance, Delaware North's dynamic family of companies has expanded to include not only sports concessions, but food service, hospitality management, retail operations, sports facility ownership and management, and gaming venue and racetrack ownership and operation.[7] The diversified organization generates over $2 billion in annual revenues, which puts it on the *Forbes* 400 list of the largest privately held companies.[8]

Delaware North worked hand-in-hand with NASA, the Smithsonian Institution, and all the contractors to design and build the Apollo/Saturn V Center at

Kennedy Space Center. The private funding for the project was provided by bus tour ticket surcharges and from an interagency agreement between Kennedy Space Center, the Spaceport Florida Authority (SFA), and South Trust Bank of Alabama. The time had come to restore the Saturn V rocket.

The Right Person to Do the Job

Finding a contractor to renovate a piece of machinery that was over a football field in length wasn't as easy as going to "R" in the yellow pages and identifying a rocket restorer. Instead, the Kennedy Space Center team sought input from some of the largest paint manufacturers in America. Among the names that kept appearing was Thomarios Painting of Akron, started in 1948 by Nick Thomarios to serve commercial and residential customers in greater Northeastern Ohio. Nick's son, Paul, expanded the business in the 1970s and 1980s, making Thomarios Painting one of the top specialty contractors in the field with the technical knowledge to solve any coating, covering, or corrosion problems.

The company was invited to bid on the project, along with some of the bigger government contractors. As the pool of candidates for the job was reduced, Thomarios remained in the running and became one of the two finalists for the project. The other contender was Bechtel, a huge multinational conglomerate that had multiple contracts with the United States government worth billions of dollars. In 1971, Bechtel had begun construction management and detailed design of the Washington, DC, Metro Transit System, and the company was a key facilitator of the 1984 Los Angeles Summer Olympics, providing plans for the conversion of over forty sports and training locations. In a true David and Goliath proceeding, Thomarios came to Florida to defeat Bechtel. Thomarios was one of many small American businessmen who understood that seizing an opportunity was not always based on quantity. The Akron entrepreneur felt a bit out of place in a room full of decision makers, but he laid out his plans and answered questions. Asked what it meant to be part of such a huge undertaking, Thomarios responded, "If we get this business, I will move down here full-time, along with several of my key employees, until the job is done. You will have us twenty-four hours a day, seven days a week."

Thomarios finished up and was asked to wait outside with the entourage from Bechtel. No one said much. Finally, a member of the group charged with announcing the winning bidder came out and graciously told the Bechtel team they could leave. He turned to Thomarios and told him the good news. His company would be the organization to restore the Saturn rocket. The reason was simple, "By giving it [the restoration work] to you, I know you will personally be accountable and act accordingly."[9]

At that moment Thomarios probably wasn't think-
ing about the Greek god, Apollo. Having won the bid,
Thomarios had to bring the decaying rocket back to
life. He would, like Apollo, be disappearing into the
darkness. He hoped the light of spring would come
soon. There were deadlines, after all.

---

1. Jo Nugent, "Smithsonian People", *The Rotarian Magazine*,
1985.
2. Clive Grout Architects, http://clivegrout.com/.
3. Carolyn Handler Miller, *Digital Storytelling: A Creator's
Guide to Interactive Entertainment* (Boston: Elsevier, 2008), p.
374.
4. Bob Rogers, interview by Andrew R. Thomas, BRC Imagination Arts, Burbank, California, March 25, 2010.
5. Ibid.
6. Ibid.
7. Delaware North Companies, http://www.delawarenorth.
com/Who-We-Are.aspx.
8. Ibid.
9. Paul Thomarios, interview by Andrew R. Thomas, July 2011.

My Story

Paul N. Thomarios

In Greek mythology, Icarus is the son of the master craftsman Daedalus. The most famous story about Icarus is his attempt to escape from Crete by using wings that his father constructed from feathers and wax. Icarus ignored the instructions not to fly too close to the sun and the melting wax caused him to fall to his death. I can relate to Icarus because I always try new things that I'm not sure will work. At the core, I'm curious.

If I didn't become a businessman, perhaps I would have become an actor. An actor prepares for a role by learning everything he/she can about a character—physical traits, speech, mannerisms, weaknesses, strengths, and other idiosyncrasies. When I prepare for a project, I get into character, too. I find out everything about not only about the job's requirements, but the people who will be involved. I engross myself in all aspects. I want to learn as much as possible— no stone unturned, no paint unmixed. I know unbridled enthusiasm might be dangerous, but no enthusiasm is worse. It's much easier to pare back than to pump up.

My story is not atypical. I learned that by working hard, with some degree of insight and know-how, I was able to get what had always seemed out of reach. My parents showed me the way.

I'm motivated by my children and my company. I try to talk to my children on a daily basis. My upbringing, a life that met my needs but wasn't extravagant in any way, provided a good perspective for what I expect of my children. I believe in their goals. I admire my daughter Sara who works full-time and will be a college graduate. I was so happy that my youngest son, Adam, married and works closely with me in the family business. Sarah may soon join us and she says, as only sisters can, that she intends to be Adam's boss one day. My oldest son, Nick, who is named after my father, is a physician, a hard-working and caring one.

Moving forward has its risk. The painting firm my father built is now moving into its third generation. When my father started painting houses, I'm sure he didn't expect his son to be restoring a Saturn V rocket. Things change. If I was certain of the future, I'd be bored. Who knows, Adam might be restoring NASA artifacts on the moon or painting houses in Akron, Ohio. There is value in both, and a lot of hard work.

Chapter 7
Bringing the Saturn V Back to Life

There is a single light of science, and to brighten it anywhere is to brighten it everywhere.
—Isaac Asimov

"She's riding like a dream." Wally Schirra, Commander, Apollo 7

"As I look down on the Earth here from so far out in space, I think I must have the feeling that the travelers in the old sailing ships used to have: going on a very long voyage away from home, and now we're headed back." Bill Anders, Lunar Module Pilot, Apollo 8

"There's that horizon. Boy, is that pretty! Wow!" Jim McDivitt, Commander, Apollo 9

"Just like old times! It's beautiful out there!" John Young, Command Module Pilot, Apollo 10

"I wonder if we could find out when the last time my lawn was cut." Buzz Aldrin, Lunar Module Pilot, Apollo 11

"We're just exercising and listening to the tape recorder and looking at the moon and looking at the Earth." Charles "Pete" Conrad, Commander, Apollo 12

"There's one whole side of the spacecraft missing." James A. Lovell, Commander, Apollo 13

"Yes it's a beautiful day here at Fra Mauro Base. Not a cloud in the sky." Alan Shepard, Commander, Apollo 14

"I got the big fireball going by at staging. I don't know whether you saw it or not. That beauty really goes." David Scott, Commander, Apollo 15

"[W]hen you look away from the Earth—or the moon—it's just the utter blackness of space. It really is black out there." Charles Duke, Lunar Module pilot, Apollo 16

"[W]e leave as we came, and, God willing we shall return, with peace and hope for all mankind." Gene A. Cernan, Commander, Apollo 17

Thomarios might have known that the first stage of the Saturn V rocket was held together by 250,000 nuts, bolts, rivets, and other fasteners in 2,000 different sizes and weights. He might have been aware that 8 million parts were used to build the entire mechanism to get men to the moon and back. It was unlikely he was familiar with the teams that worked on the original rockets, with names like space coroners, scientific Vic Tannys, rocket surgeons, tailors, and buyers, and who had built and tested the Saturn V. Space coroners autopsied the engines after flight to detect signs of foul metal play or other suspicious parts activity; scientific Vic Tannys, named after the once famous body builder and health club pioneer, tried to find ways to keep unwanted pounds off the structure; rocket surgeons analyzed parts in rooms cleaner than operating theaters; tailors had to build form-fitting wardrobes for the rocket's exterior; and buyers went shopping for items from diapers to empty beer cans. The cost to launch each Apollo mission was around $5 billion in today's dollars.

The stages of the Saturn rocket were, for all intents and purposes, large cylinders that held engines and fuel to power the engines. The first stage was driven by five Rocketdyne F-1 engines and operated for less than 180 seconds—long enough to lift the rocket to an altitude of approximately 38 miles. On the Apollo 11 flight, Neil Armstrong signaled "inboard cut-off" at 2 minutes, 17 seconds into the launch. The 300,000-pound stage, 33 feet in diameter and 138 feet long, fell back into the Atlantic Ocean shortly thereafter, having burned up more than 4.5 million pounds of propellant. The center F-1 engine was fixed and the outer four engines swiveled, or gimbaled, for steering control. Fuel was pumped to each engine at a speed of approximately 200,000 gallons/minute producing 1.5 million pounds of thrust.

On the Apollo 7 mission, the NASA Public Affairs Officer noted two events right after the rocket's staging commenced. "He's [referring to flight commander, Wally Schirra] got ignition and he says we are up to thrust on the second stage. The thrust is okay at 2 minutes 40 seconds into the flight."[1] The second stage's propulsion came from five Rocketdyne J-2 engines, which some have called the most important engine in the history of manned space flight propulsion. The J-2 used a cryogenic propellant (liquid oxygen to oxidize the liquid hydrogen fuel) and was designed to be restarted multiple times during a mission. The fuel combination had to be kept at very low temperatures or the mixture would return to a gaseous state. Since the cryogenic fuel yielded more power per gallon, a given mission could be accomplished with a smaller quantity of fuel and a smaller vehicle, or the mission could be accomplished with a larger payload than possible with the same mass of conventional propellants.

The 81-foot-long and 33-foot-wide second stage fired for about six minutes. Each J-2 was capable of creating of 225,000 pounds of thrust. At 7 minutes and 42 seconds into the Apollo 11 launch, the second stage engines were shutdown. After engine cutoff and separation, the second stage plunged into the Atlantic Ocean about 2,400 miles from Kennedy Space Center.

At 9 minutes 52 seconds into the Apollo 16 launch, mission commander, John Young acknowledged that the stage II engines had been shutdown. Five seconds later, he informed mission control "We have S-IVB [third stage] ignition."[2] At 11 minutes and 49 seconds into Apollo 16's ascent, Ken Mattingly, the command module pilot radioed that "SECO," S-IVB engine cutoff, had occurred and Apollo 16 was in Earth orbit. The 58-foot-long, 28-foot-wide third stage contained the Apollo payload—the command, service, and lunar modules.

All the Apollo crews made multiple checks of their craft at this point, before the initiation of the precisely-timed translunar injection procedure which reignited the third stage's J-2 engine. The firing provided enough velocity (25,000 mph) to escape the Earth's gravitational pull and propel the ship towards the moon. At 2 hours 44 minutes and 15 seconds of Apollo 11's flight, the crew began the 6 minute 44 second engine burn to put them on the path for their historic moon landing.

As the moon missions' spacecrafts drifted towards their lunar orbits, their crews needed to make one additional maneuver in preparation for landing. The command and service modules (CSM) were detached from the lunar module (LM). The CSM was rotated 180 degrees and docked nose-to-nose with the LM. For the Apollo 11 flight, the separation of the CSM took place at 3 hours 16 minutes and 59 seconds into the mission, and after a period when communication was lost with the Houston Space Center, Neil Armstrong confirmed at 3 hours 29 minutes and 35 seconds that the CSM and LM had docked. After some additional equipment checks were performed, the last of the Saturn stages was jettisoned, having successfully propelled the astronauts to the moon.

The Apollo 13 explosion and rupture of one of the service module oxygen tanks probably focused more attention on the command, service, and lunar modules than even Apollo 11's groundbreaking flight. Fifty-six hours into the Apollo 13 flight, the possibility that the three-man crew would be stranded in space was a situation with which NASA was grappling. Perhaps, the agency should have recognized what many high-rise developers had known for years—skip the number 13 altogether and go from the 12th to the 14th floor. Luckily, the crew of Apollo 13 returned to Earth safely, and the mishap highlighted the design and engineering sophistication of the CSM and LM.

Command Module

The Apollo Command Module (ACM) was a descendent of the Mercury and Gemini capsules, all three of which were designed by NASA's chief engineer, Max Faget. The cone-shaped command module had a diameter of 12 feet, 10 inches, a height of 10 feet, 7 inches and weighed about 13,000 pounds. The quarters for a three-person crew were cramped, to say the least.

After the Apollo 1 launch pad fire which killed three astronauts, extra safety features were incorporated into the ACM design. The pure oxygen environment was changed to a less flammable oxygen/nitrogen mix. An escape lever, situated above the center astronaut, opened the hatch and broke the pressure seal simultaneously, allowing for a 15–25 second escape in an emergency situation, a great improvement over the 90 second unlock procedure on Apollo 1. Flame-retardant materials were used in the interior capsule construction and the astronauts' suits were enhanced to be more heat-resistant. The ACM's forward section contained two reaction control system engines and docking probe and tunnel. The command module featured two observation windows, two rendezvous windows, and one hatch window. The aft compartment contained ten reaction control system engines and their propellant tanks, fresh water tanks, and umbilical cables which linked the command module with the service module. Since the ACM was the only piece of equipment from the enormous moon rocket to return to Earth, the capsule was equipped with a heat shield to enable re-entry and parachutes to facilitate a controlled splashdown.

Service Module

The Apollo Service Module (ASM) provided the basic propulsion and maneuvering systems for the Apollo missions after leaving Earth's orbit. The cylindrical service module measured 24 feet, 7 inches in length, and 12 feet, 10 inches in diameter, and weighed almost 54,000 pounds. The ASM had all the communication and life-sustaining equipment the crew needed in space, including the fuel cell power system, cryogenic hydrogen and oxygen tanks, as well as helium and oxidizer tanks.

The key feature of the ASM was the Service Propulsion System engine. The engine could produce 20,000 pounds of thrust using hypergolic fuels, which combust on contact without the need of an igniter. The engine not only helped achieve lunar orbit, but had to provide the critical burns to put the crew on a path back to Earth. On the Apollo 11 mission, the lunar module's engines fired to achieve the descent orbit injection around the moon, a procedure that limited the hovering capability of the lunar module and left Apollo 11 with only 30 seconds of fuel before touching down on the lunar surface. The ill-fated Apollo 13 service module was to perform this function, providing a cushion for the lunar landing module. Obviously, this modification went out the window when the service module became little more than deadweight on that mission.

Lunar Module

After NASA decided that a direct-ascent method (launching a rocket to the moon, landing it on the surface, and returning it to Earth) was not feasible, engineers and designers began to work on a lunar rendezvous method. Under that scenario, a smaller craft was needed to undock from the mother ship and land on the lunar surface. One astronaut called what was ultimately created a "toaster oven with legs." Another described the craft as an "ugly and unearthly bug." Some named it the "Spider" or the "Lunar Schooner."

Whatever the term, the Apollo Lunar Module was a two-stage vehicle designed to carry astronauts from a lunar orbit to the surface of the moon and then back to rendezvous with the command and service modules. The entire ship was 22 feet, 11 inches high, and 14 feet, 1 inch wide, and weighed about 8,500 pounds. Its unique look was a result of the fact that the ship was only to be used in space and was freed from many design requirements of Earth's gravity.

The lunar module consisted of two separate sections. The descent stage housed the fuel and engine necessary for a controlled, powered decent to the lunar surface. This stage also contained the landing gear, which cushioned the lunar module as it landed on the surface and supported the vehicle upright, the equipment astronauts used to explore the moon, and television equipment to send images to Earth. The descent stage also became the launch pad for the return trip to the command module. The ascent stage had room for two astronauts. The stage had environmental controls, an engine, fuel, and an electrical power supply. The stage was equipped with a docking tunnel hatch to enable the crew to get to and from the command module and a hatch to get to the surface of the moon. The exterior had reaction control engines, with which to maneuver, and communications apparatus.

A little over 100 hours into the Apollo 11 flight, the LM separated from the command module for its descent to the moon's surface. Michael Collins, in the command module, told the LM's passengers, Neil Armstrong and Buzz Aldrin, "Okay, there you go. Beautiful!" Commander Alfred Worden of Apollo 15 welcomed back his lunar voyagers, David Scott and James Irwin, at about 173 hours and 37 minutes into their mission. Scott and Irwin were "glad to be home."[3]

The Final Climb

The Apollo program was a great engineering feat. Ironically, the end of the program also left many engineers without jobs. Technology changes quickly. Downsizing is a fact of life. Industries have to reinvent themselves constantly. Science benefitted in several ways from the moon exploration, but no extraordinary discovery was impactful enough to spur the imagination of funders. Other manned interplanetary travel seemed less exciting. Perhaps, the idea that only a small capsule was returned after an Apollo flight just didn't fit with a new era of sustainability. The historical record ebbs and flows. As one generation's accomplishments become chapters in elementary school texts, important dates hold less relevance. Museums are key connectors to the past. We love the "what ifs." Our fascination with history might be part of our genetic makeup.

Paul Thomarios kept his word and moved first to a hotel in the area and then into an apartment at Cape Canaveral. He stayed on site during the week and flew home on weekends to keep tabs on the other projects that were in progress. The lively Thomarios racked up frequent flyer miles and when he stayed on at the Cape one weekend, the flight crew asked if anything bad had happened when he boarded the plane to Cleveland the following week. Being raised in a gritty Midwestern city was a vital part in the development of Thomarios' work ethic. As he was fond of saying, "If you can make it in Akron, Ohio, you can make it anywhere." Apologies to the Big Apple.

The accelerated time frame didn't allow Thomarios to hire a local crew. The rocket needed to be restored by May 15, 1996. Bringing his Ohio employees to the

Florida coast was the most prudent option. During the restoration process, workers lived in rented houses and apartments or in local hotels. To ensure some sense of stability, wives were able to visit on weekends. The work schedule, three eight-hour shifts, seven days a week, for the duration of the project, precluded employees from returning to Ohio. Thomarios just couldn't have any disruptions to the schedule.

Initially, Thomarios broke his crew into separate teams to work on different projects. To abate boredom, he rotated the team members from one team to another. Eventually, individuals developed specific skills like rewiring, metal working, or coating, and more permanent teams were formed. Thomarios continually worked to motivate the teams. In many instances, the tight deadlines caused friction and misunderstandings. Thomarios worked to mollify the individuals and keep them focused on the final goal that could only be reached through cooperation and group accomplishments.

The first priority was to get the rocket parts away from the elements and into an environmentally-friendly work space. Thomarios set up an air-supported structure in the parking lot next to the rocket display and the structure served as the workshop for the second and third stages. Originally designed to cover tennis courts, the air structure provided cover and protection for much of the restoration and preservation work.

The massive first stage was too big to fit into the air structure. Something else was required, so scaffolding was built around the entire stage. This gave the crew complete access to all of the components, including the massive F-1 engines and the rear heat shield. Composed of asbestos, the panels of the heat shield prevented the aluminum housing of the rocket from melting during launch.

The rocket's bulk was one of many obstacles that hindered the restoration process. All restoration had to be carefully documented; when weathered parts were replaced or rebuilt completely, the Smithsonian Institute wanted this noted. The final display would indicate to visitors which parts were original. Safety rules also kept changing. NASA had become a risk-averse operation, especially after the *Challenger* disaster. Many hours were spent consulting with NASA folks to ensure that nothing was being missed during the restoration. While important, the meetings, one of which included seventy-three NASA officials and ten people from the Smithsonian, took up valuable time needed to finish the job.

After twenty years in the Florida elements, the Saturn rocket needed a thorough scrubbing. The rocket parts were covered with mildew, chewing gum, bird feces, and other items that defied description, but stuck to the rocket's exterior. The gunk was so thick that Nick Bolea, a long-time Thomarios employee, decided to get on his employer's good side by using a power washer to write "Thomarios" in five-foot-high letters on the rocket's exterior.

The restoration crew began to wash away the layers of grime. Due to NASA regulations, pure, deionized water had to be used. Official were worried that any heavy metals in other water could harm the rocket's exterior and were concerned that runoff would impact the local water table. Even though 30,000 gallons of water evaporated during the operation, the Thomarios crew disposed of an additional 10,000 gallons at a certified site. While large parts of the exterior were cleaned using 3,000 psi blasts, other parts of the rocket's exterior were simply too fragile and a good deal of elbow grease had to be applied.

The Ohio boys stopped work one day because a mysterious purple runoff was oozing out of the rocket. A hazardous material team was summoned to investigate. After some analysis, the team discovered the material wasn't dangerous after all, but was fruit juice from berries birds had stored in the rocket's interior. NASA engineers undoubtedly hadn't designed the rocket for this kind of life support, or for the eight-foot tall tree that had to be cut out of the interior near the instrument panel.

Northerners usually find out about alligators after somebody misplaces one down a sewer system and a camera crew shows the little creature on the six-o'clock news. Florida has real gators and they don't just play football on fall Saturdays. The Ohio crew befriended George, a local alligator that often toured around the parking lot to see what was happening. Told not to feed the creature, the crew did anyway. At first, the relationship was symbiotic. George got fed and Thomarios's workers pretended they were animal trainers.

Gators are very quick and can easily overtake prey trying to escape. Alligators are like Pavlov's dogs, too. While alligators don't usually associate people as food, after many feedings, George changed his mind. One day when George was deprived of his lunch, the gator started looking for sustenance in all the wrong places and the crew had to scramble to safety. Fortunately, no one got hurt or eaten, and the local game preserve folks came by and relocated George down the coast to Merritt Island, Florida.

After many layers of surface filth were washed away, the crew found up to eight layers of paint had been applied to keep the rocket from totally disintegrating in the Florida elements. Thomarios experts used a combination of two techniques to advance the restoration. First, the team blasted away old paint and light corrosion with fine grades of ordinary baking soda, a technique capable of stripping paint without etching the metal. The baking soda was rinsed to remove its salt content prior to its application. The machine used to do the work, however, reverberated like a dentist's ultrasonic plaque-removal tool, adding to the constant din under the dome. The baking-soda procedure worked beautifully on loose paint, but couldn't handle the well-adhered paint and corrosion. For the stubborn layers, Thomarios used a removal technique with volcanic sand as the abrasive.

A museum director once noted because the Saturn V rockets had been neglected for an extended period of time, "You can't just do little touch-ups. It has to be done right."[4] The removal steps revealed many other issues. Prior to treating the first stage, a worker had made a knife hole in the 3/16-inch-thick aluminum to test its corrosion depth. The wear didn't seem that extensive. However, the abrasive blasting revealed massive degradation, and when the area was finished, the small hole had become a crater in which three men could stand.

A second issue involved the polyurethane foam. This material was sprayed on to ensure that the supercooled liquid hydrogen would not boil off prior to launch. Birds had pecked numerous holes into the surface. These had to be fixed.

While the rocket sat outside and uncovered, rain water seeped between the insulation and the aluminum. As the water evaporated, the trapped vapor caused almost twenty blisters to form in the insulation. First, workers reinforced the damaged areas with new metal stringers, returning them to their original appearance. Next, each area had to be painstakingly cut out with a knife and filled with automotive-type body filler. When the filler dried, the areas were finally sanded to replicate the original fuselage contours. Work conditions inside the bubble were "nasty" according to John Lilly, a burly but gentle man known on the site for being able to fix anything. The space was dense with baking soda dust and the air was soggy. Electrical machinery clogged and broke down. Workers sweated profusely. Hammers pounded. Grinders screeched, pneumatic guns pinged and popped, compressors hissed, and industrial fans constantly whirred to keep the balloon-like building inflated.

Mike Ciocca, a plasterer, found the outside conditions little better. The mean high temperature during the project was around eighty degrees. Of the three-and-a-half months on the job, the skies had little or no cloud cover during sixty-eight days, and the maximum humidity reached over 50 percent on forty-four days. Ciocca had to replace the original heat shields because they contained asbestos and did not meet EPA standards. Each of the eighty shields was shaped differently, requiring re-fabrication with a plywood base topped with Styrofoam, nylon mesh, and synthetic stucco.

In the end, a good number of the aluminum panels, which were originally designed and fabricated using knowledge gained from airplane manufacturing, had to be re-fabricated. The original panels were made of different aluminum alloys, chosen due to their welding features, strength under cryogenic conditions, and resistance to stress corrosion. Norm Wilson, who was the manager of one of the Saturn's manufacturing engineering sections, recalled that you "can't really say our work has been exotic. But when you consider the sizes, angles, lengths, designs, offset tolerances, and overall specifications involved, you have one challenging welding problem on your hands. We've had to tap our experience well dry and tax our imagination to come up with the right answers, and it has been only through the combined contributions of many that we have been successful."[5] The Thomarios men understood Wilson's predicament during the twenty-thousand man-hours of work at the Cape.

One last item had to be dealt with prior to completing the job. The size of the American flags on the rocket's first stage could not be determined. Mold and birds had completely eaten them away. Thomarios sought advice and hired a vexillologist. Even though the name may sound like a character in a horror film, vexillology is the study of flags. After examining pictures of the original moon rockets, the flag expert determined the height and width of the flags, and the flags were repainted to proportion on the rocket.

There is a bit of irony in this last step. In April 1969, two members of the North American Vexillological Association (NAVA) wrote a paper on the appropriateness of planting a flag on the moon's surface during the Apollo 11 flight. The two researchers recognized "that there is no atmosphere on the moon and that an ordinary cloth flag would not display itself. It would, however, be possible to erect an inflexible flag made of some light but rigid composition, with the colors permanently inlaid in fluorescent dyes." They decided that while there was some precedent for displaying a national marker for future generations, "a recent international treaty [had] restricted claims to sovereignty on the moon and the planets."[6] The NAVA folks considered it more fitting to put up a plaque indicating that the quest to the moon was an act of humankind. NASA's wishes trumped those of NAVA.

Thomarios's crew had to perform their work around the space shuttle's schedule as well. On January 11, the space shuttle *Endeavor* was launched and landed at the Cape eight days later. February 22 saw the launch of *Columbia*. On March 22, *Atlantis* departed. And May 19 saw *Endeavor* takeoff on another mission.

The intensive preparation for each launch brought heavy activity to the area around the Vehicle Assembly Building. Launch vehicles had to be transported at a crawl to the launch pad. Launches themselves limited or precluded restoration work. Thomarios was always busy adjusting schedules, reworking plans, and finding time slots to finish projects. The day started whenever the crew could be on site and ended at odd intervals. When May 15 dawned, the work had been completed and the restored Saturn rocket could be moved to become part of the museum experience at the Kennedy Space Center. The center itself cost about $35 million. The hundred-thousand-square-foot attraction has four components: The Firing Room Theatre, Saturn V Rocket Plaza, Lunar Space Theatre, and New Frontiers Gallery. The Firing Room Theatre contains the actual equipment in the Apollo firing room, accurate down to the rotary dial phones, tattered chairs, half full ash trays and half spent pack of Chesterfields and Old Gold cigarettes. However, the Saturn V Rocket Plaza is the heart of the pavilion. The three-stage rocket lies horizontally, stretching as long as a football field. It is suspended about fifteen feet off the concrete floor, giving visitors a close-up look at history. Apollo mission artifacts, including the command module, a rebuilt lunar module, and a one-ton cleat from the steel-tracked crawler that took the Saturn V from the Vehicle Assembly to the launch pad, are distributed throughout the Saturn V exhibit.

Initially, Thomarios had viewed the remnants of the Apollo program as rusty pieces of machinery. Since he was in the business of repainting and restoring, the project fit within his company's expertise. As the project continued, he became aware of the large numbers of people who were involved with the

Apollo program and the gratitude they were expressing with the progress of the restoration work. It really hit home, when the commander of Apollo 12, Charles "Pete" Conrad wanted to meet him. The successful completion of the Saturn V restoration had other positive effects. Paul and his company became known as the rocket guys. In subsequent years, Thomarios restored and preserved the very first Mercury capsule, an original Atlas/Mercury rocket, and an original Titan/Gemini rocket and capsule for the New York Hall of Science; a boilerplate Apollo capsule for the Detroit Science Center; and moved the Apollo/Soyuz capsule to the Great Lakes Science Center in Cleveland, Ohio.

His business also expanded after the Saturn success. The knowledge his workers gained at Cape Kennedy allowed the company to branch out further into areas like metalworking and industrial painting. The exposure also brought new contacts, including other astronauts, politicians, inventors, and innovators.

Orville Wright mused if "we worked on the assumption that what is accepted as true really is true, then there would be little hope for advance." Charles Lindbergh asserted "... no one can tell just how far flying will take us." Amelia Earhart realized "... the lure of flying is the lure of beauty." Charles Yeager understood "... if you don't have any control over the outcome of a flight, don't worry." Risk comes with every advance.

In retrospect, only sixty-six years passed from Orville and Wilbur Wright's successful airplane flight until the first human step onto the moon's surface. Only forty-three years went by from the time Goddard launched his first liquid-propelled rocket before Neil Armstrong set foot on the lunar surface. And remarkably, only eight years passed after Yuri Gagarin became the first man to orbit Earth prior to the Eagle landing on the moon.

While there is some dispute about the exact date for the start of human evolution, most researchers believe the event happened between one hundred thousand and two hundred thousand years ago. The time from Kitty Hawk to Tranquility base represents a mere ³⁄₁₀₀ths of a second along man's timeline. The Thomarios project to restore the Saturn rocket lasted only six months, not even a blip in time. Still, the small segment of human existence might provide what Thomarios sees as the greatest benefit—"inspiring kids and future generations." As he is fond of saying, "We have to find the next big thing. We need to arouse the imagination of coming generations. I was nothing more than a 'C' student. Maybe that helped because I didn't know what to do. I didn't know how to say 'no.' I was probably too dumb to understand all the consequences. I'm not that cautious. Hard work overcomes a lot of shortcomings."[7]

1. Smithsonian Institution, National Air and Space Museum, Apollo 7, http://www.nasm.si.edu/collections/imagery/apollo/as07/a07.htm.

2. Smithsonian Institution, National Air and Space Museum, http://www.nasm.si.edu/collections/imagery/apollo/AS16/a16.htm.

3. Smithsonian Institution, National Air and Space Museum, http://www.nasm.si.edu/collections/imagery/apollo/AS11/a11.htm; Smithsonian Institution, National Air and Space Museum, http://www.nasm.si.edu/collections/imagery/apollo/AS15/a15.htm.

4. Tamer El-Ghobashy, "Hall of Science rockets 'take off,'" *New York Daily News*, August 12, 2001.

5. Bilstein, *Stages to Saturn*.

6. *NAVA News*, 1969, Vol. 2, no. 3.

7. Paul Thomarios, interview by Andrew R. Thomas, July 2011.

Afterword

Neil Armstrong is an icon—the first man to set foot on the moon. The mission of Apollo 11 holds a weighty status, not unlike the first heart transplant. Every decade, the story of the Eagle landing on the moon's surface is feted with another anniversary. When people think of the fruition of John F. Kennedy's promise, the launch and splashdown of astronauts Armstrong, Aldrin, and Collins come to mind.

Apollo 13 is remembered today due to the success of Tom Hank's movie of the same name. Other moon landings are recognized for a golf shot or the bumpy ride of lunar buggies or a bag of moon rocks. But Apollo 12 should hold a status all its own and not as the forgotten middle child.

Apollo 12 stands as the greatest testament to the resiliency and robustness of the Saturn V. Apollo 12 is a composite of the vast ingenuity and innovation that drove the space program. Apollo 12 encapsulated America's reasons for originally embarking on the quest to explore space.

When Apollo 12 was launched in late 1969, a Ford Maverick cost $1,995. The new Polaroid Colorpack II was $29.95 and a perfect holiday gift. Long-playing records were all the rage, and joining the Columbia House Album of the Month Club came with a first-time selection of 12 long plays for $3.98, plus shipping and handling. Today, a few Mavericks might have historical license plates on them. LPs have been replaced by cassettes, then CDs, and now digital downloads. Phones come with cameras and images can be instantly uploaded to social web sites. Technology has transformed all aspects of life, except perhaps the human spirit.

On November 14, 1969, the Saturn V, carrying astronauts Charles 'Pete' Conrad, Richard Gordon, and Alan Bean, lifted off successfully. Within seconds of liftoff the Saturn spacecraft was struck by lightning. Seconds later, the massive structure was struck again. As Conrad put it during his debriefing report:

> I knew that we were in the clouds; and, although I was watching the gauges I was aware of a white light. The next thing I noted was that I heard the Master

Alarm ringing in my ears and I glanced over to the caution and warning panel and it was a sight to behold....I guess the most serious thing was the second lightning strike, which we weren't aware of. I was under the impression we lost the platform simply because of low voltage but apparently that's not the case. Apparently we got hit a second time at that point...All we needed to do was blow a battery off the line and I have a decided impression we would have gotten an Auto abort.[1]

At Mission Control, engineers struggled to understand what had happened. Neither flight director Gerry Griffin nor the senior controllers had ever experienced such a situation and contemplated scrubbing the mission. The crew would have had to jettison their capsule using the launch escape rockets mounted atop the command module. Once the crew had cleared the rocket, the mighty Saturn V would have been destroyed by mission control using explosives, and the Apollo 12 mission would have resulted in a tremendous monetary and public relations disaster.

Instead, Griffin turned to the 24-year-old engineer named John Aaron, who oversaw the electrical system on the mission, for advice. Even though Aaron's display was a meaningless jumble of numbers, he had a sudden inspiration. He remembered a similar problem during training simulations a year earlier and how that had been resolved.

Over the communication system, Aaron gave the command, "Flight, try SCE [Signal Condition Equipment] to Aux." Pete Conrad, misunderstanding, gave a pointed response. "Try FCE to Auxiliary. What the hell is that?"[2] Luckily, Alan Bean remembered that the SCE switch was located near his seat and followed Aaron's instructions. Moments later, the flight telemetry was restored in Mission Control and the ground engineers knew the vehicle was still operating properly.

Although the electrical and navigation systems within the command module had been affected by the lightning strikes, the guidance system on the Saturn V had continued to function perfectly. This system kept the rocket on its proper path into Earth orbit and the mission was never in any actual danger. The crew managed to get the rocket's fuel cells back up after second-stage engine firing and the command module's inertial guidance system was realigned once in orbit. Apollo 12 entered its planned Earth orbit 11 minutes and 44 seconds after liftoff. All systems were rechecked and the crew was given the go-ahead to head for the lunar surface.

The primary mission of Apollo 12 was to retrieve components from Surveyor 3. Launched in April 1967, Surveyor 3 carried a survey television camera, as well as other instrumentation for determining various properties of the lunar surface material, and a surface sampler instrument for digging trenches, making bearing tests, and otherwise manipulating the lunar material in view of the television system. During its operation, which ended May 4, 1967, Surveyor 3 acquired a large volume of new data and took 6,326 pictures. In addition, the surface sampler accumulated 18 hours of operation, which yielded significant new information on the strength, texture, and structure of the lunar material to a depth of 17.5 centimeters.

The Apollo 12 astronauts landed within 500 feet of Surveyor 3 and brought the camera back with them. The accuracy with which the mission was carried out, after two lightning strikes, was phenomenal. The Saturn V proved its design and strength capabilities.

In the twelve years since the launch of Sputnik, the United States had become the global leader in astronautic technology. Precision guidance systems allowed the US to create large advances in military capability, which fashioned Soviet policies and outlook. Perhaps too, the understanding that the Earth was a small part of a greater universe impacted the coming détente between the US and the USSR.

Launch pads have been replaced by museum displays and heady talk of discovering inhabitants on other planets have given way to video games. The space program has gone the way of Pluto. That's the reality, but so is the fact that the Saturn V took America—and all of humankind—on its greatest journey. The quest to put a man on the moon uncovered a very fundamental truth about humanity. Drive and passion allow the impossible to become real. In the end, as Thomas Edison put it, "there is no substitution for hard work."

---

1. Smithsonian Institution, National Air & Space Museum, http://www.nasm.si.edu/collections/imagery/apollo/AS12/a12.htm.
2. Ibid.